LANG—Abelian Varieties
LANG—Diophantine Geometry
LEFSCHETZ—Differential Equations: Geometric Theory, 2nd Edition
LOVELOCK and RUND—Tensors, Differential Forms, and Variational Principles
MAGNUS, KARRASS, and SOLITAR—Combinatorial Group Theory
MANN—Addition Theorems: The Addition Theorems of Group Theory and Number Theory
MARTIN—Nonlinear Operators and Differential Equations in Banach Spaces
MELZAK—Companion to Concrete Mathematics
MELZAK—Companion to Concrete Mathematics, Volume II
MEYER—Introduction to Mathematical Fluid Dynamics
MONTGOMERY and ZIPPIN—Topological Transformation Groups
MORSE—Variational Analysis: Critical Extremals and Sturmian Extensions
NAYFEH—Perturbation Methods
NIVEN—Diophantine Approximations
ODEN and REDDY—An Introduction to the Mathematical Theory of Finite Elements
PLEMELJ—Problems in the Sense of Riemann and Klein
PRENTER—Splines and Variational Methods
RIBENBOIM—Algebraic Numbers
RIBENBOIM—Rings and Modules
RICHTMYER and MORTON—Difference Methods for Initial-Value Problems, 2nd Edition
RIVLIN—The Chebyshev Polynomials
RUDIN—Fourier Analysis on Groups
SAMELSON—An Introductions to Linear Algebra
SANSONE—Orthogonal Functions
SIEGEL—Topics in Complex Function Theory
Volume 1—Elliptic Functions and Uniformization Theory
Volume 2—Automorphic Functions and Abelian Integrals
Volume 3—Abelian Functions and Modular Functions of Several Variables
STOKER—Differential Geometry
STOKER—Nonlinear Vibrations in Mechanical and Electrical Systems
STOKER—Water Waves
TRICOMI—Integral Equations
WASOW—Asymptotic Expansions for Ordinary Differential Equations
WHITHAM—Linear and Nonlinear Waves
YOSIDA—Lectures on Differential and Integral Equations
ZEMANIAN—Generalized Integral Transformations

APPLIED NONSTANDARD ANALYSIS

APPLIED NONSTANDARD ANALYSIS

MARTIN DAVIS

Courant Institute of Mathematical Sciences
New York University

A WILEY-INTERSCIENCE PUBLICATION

JOHN WILEY & SONS New York · London · Sydney · Toronto

Library of Congress Cataloging in Publication Data

Davis, Martin, 1928-
Applied nonstandard analysis.

(Pure and applied mathematics)
"A Wiley-Interscience publication."
Bibliography: p.
Includes index.
1. Mathematical analysis, Nonstandard. I. Title.

QA299.82.D38 515 76-28484
ISBN 0-471-19897-8

Printed in the United States of America

10 9 8 7 6 5 4 3 2 1

To the Memory of Abraham Robinson

PREFACE

Nonstandard analysis employs the methods of model theory (itself one of the important branches of mathematical logic) to embed the universe of mathematical discourse in a so-called "nonstandard universe" containing infinitesimals as well as infinite objects. It is the purpose of this book to make the methods of nonstandard analysis available to readers with no previous knowledge of logic. The emphasis throughout is on the *applications* of nonstandard analysis to "standard" mathematics (as opposed to investigations of the nonstandard universe in and for itself). Since nonstandard analysis is of potential interest to readers with varied backgrounds, the prerequisites have been kept quite modest. For the first three chapters, knowledge of the simplest properties of groups, rings, and fields and some previous exposure to $\varepsilon - \delta$ methods should suffice. For the last two chapters, some knowledge of the elements of finite dimensional linear algebra is also assumed.

The emphasis in the first three chapters is on natural intuitively suggestive proofs of results whose standard proofs are more complicated. In the last two chapters, standard and nonstandard methods are freely used as convenience dictates in obtaining results about linear operators, assuming no previous knowledge of functional analysis. In particular, the Bernstein–Robinson proof of the existence of invariant subspaces for a polynomially compact operator on a Hilbert space is included.

This book developed from graduate courses given at the Courant Institute and at the University of British Columbia, and it is intended to be usable as a textbook for an advanced undergraduate or graduate course. I have profited in various ways from students and colleagues who participated in my courses. I have been particularly helped by comments of Peter Gilkey, Melvin Hausner, and David Russinoff. This book would never have been written without the class notes prepared by Barry Jacobs. Finally, it is a pleasure to acknowledge with gratitude that Connie Engle's typing of the manuscript was up to her usual impeccable standard.

Nonstandard analysis is very much the creation of one man, Abraham Robinson, whose recent untimely death has been a great loss to mathematics. The reader may assume that results not otherwise attributed are his.

It is a great historical irony that the very methods of mathematical logic that developed (at least in part) out of the drive toward absolute rigor in analysis have provided what is necessary to justify the once disreputable method of infinitesimals. Perhaps indeed, enthusiasm for nonstandard methods is not unrelated to the well-known pleasures of the illicit. But far more, this enthusiasm is the result of the mathematical simplicity, elegance, and beauty of these methods and their far-reaching application.

MARTIN DAVIS

New York, New York
November 1976

CONTENTS

GLOSSARY OF SPECIAL SYMBOLS

Symbol	Meaning	Page
$\mathscr{P}(A)$	Power set of A	5
$\langle a,b\rangle$	Ordered pair of a and b	5
$\langle x_1,x_2,\ldots,x_n\rangle$	Ordered n-tuple	6
$g[C]$	Image of C under g	6
N	Set of natural numbers	6
N^+	Set of positive integers	6
$\bigcup_{i\in I} X_i$	Union of an indexed family of sets	7
$\bigcap_{i\in I} X_i$	Intersection of an indexed family of sets	7
$\prod_{i\in I} X_i$	Cartesian product of an indexed family of sets	7
μ_F	The measure induced by the ultrafilter F	10
$\hat{S}$	The superstructure with individuals S	12
$r \upharpoonright s$	Equals $r(s)$, when this latter is defined	15
a.e.	Almost everywhere	16
$\mathscr{L}_U$	The language corresponding to universe U	20, 21
$\lvert\mu\rvert_U$	The value of closed term μ of $\mathscr{L}_U$ in U	22
$U \models \alpha$	The sentence α is true in U	23
$U \not\models \alpha$	The sentence α is false in U	23
$\mathscr{L}$	The language corresponding to the standard universe	24
$^*\mathscr{L}$	The language corresponding to the nonstandard universe	24
$^*\lambda$	The image of a term or formula of $\mathscr{L}$ in $^*\mathscr{L}$	24
$^*A$	Subset of nonstandard universe corresponding to definable subset A of the standard universe	29, 30
Q	The ordered field of rational numbers	44
F,I	The sets of finite and of infinitesimal elements, respectively: of an ordered field	47
	of the hyperreal numbers	55
$x \approx y$	x is infinitesimally near y	47, 76, 89

Symbol	Meaning	Page
$^\circ x$	The image of $x \in F$ in F/I under the natural homomorphism	48
	The standard part of hyperreal x	55
	The standard part of a near-standard point	89
R	The ordered field of real numbers	53, 54
R^+	The set of positive real numbers	56
$[a,b]$	The closed interval with endpoints a,b	57
$s_n(x) \rightrightarrows s(x)$	$s_n(x)$ converges uniformly to s(x)	62
$f \sim g$	Local maps f,g are equivalent	63, 117
$\mathcal{O}$	Collection of open sets of a topolgical space	75
$\mathcal{O}_p$	Collection of open neighborhoods of p	76
$\mu(p)$	The monad of p	76
$\rho(x,y)$	The distance between x and y in a metric space	87
$B_p(r)$	The open ball with center p and radius r	88
$\|x\|$	Norm of x	105
$\mathscr{C}[\mathscr{B}]$	Space of continuous maps from a real interval into Banach space $\mathscr{B}$	111
$\mathscr{S}$	The normed linear space of step functions on a real interval	111
(a,b)	Inner product of a and b in a unitary space	121
δ^n_m	Kronecker's delta	122
$a \perp b$	a is perpendicular to b	122
$T_1 \leq T_2$	The ordering of Hermitian operators on a unitary space	122
l^2	Hilbert space of infinite sequences of complex numbers	123
P_n	Projection of l^2 into an n-dimensional subspace	129
span$(t_1, \ldots, t_k)$	Linear space spanned by $t_1, \ldots, t_k$	131
P_E	The projection operator on closed linear subspace E	132
$E^\perp$	The orthogonal complement of closed linear subspace E	132
$A \perp B$	Subspaces A,B are orthogonal	134
$G = E \oplus F$	Direct sum of orthogonal closed subspaces E,F	134
W	Set of subscripts corresponding to infinite eigenvalues	153

APPLIED NONSTANDARD ANALYSIS

INTRODUCTION

1. WHY NONSTANDARD ANALYSIS?

Nonstandard analysis is a technique rather than a subject. Aside from theorems that tell us that nonstandard notions are equivalent to corresponding standard notions, all the results we obtain can be proved by standard methods. Therefore, the subject can only be claimed to be of importance insofar as it leads to simpler, more accessible expositions, or (more important) to mathematical discoveries.

As to the first, the reader must be the judge. The best evidence for the second is the Bernstein–Robinson theory of invariant subspaces of infinite dimensional linear spaces, which settled a question that had remained open for many years. Quite simple standard proofs of their results now exist. Nevertheless, we develop part of their theory not only because this was the path of discovery, but also because it gives us an opportunity to exhibit the truly beautiful idea of approximating an infinite dimensional space from *above* by a space to which the results of *finite* dimensional linear algebra are applicable.

2. INFINITESIMALS AS IDEAL ELEMENTS

The study of nonstandard analysis locates itself, mathematically, in the context of the method of ideal elements. This is a time-honored and significant mathematical idea. One simplifies the theory of certain mathematical objects by assuming the existence of additional "ideal" objects as well. Examples are the embedding of algebraic integers in ideals, the construction of the complex number system, and the introduction of points at infinity in projective geometry.

Nonstandard analysis involves introducing ideal elements that are infinitely close to the objects we are interested in and also ideal elements that are infinitely far away. Leibniz first made use of these ideas in developing differential and integral calculus. In spite of the fact that this was intuitively a very appealing way to think, it seemed to be impossible to provide a sound mathematical basis for the theory of infinitesimals.

There is no difficulty with the purely algebraic problem of embedding the real numbers in a field containing infinitesimals (i.e., what is called in algebraic language a non-Archimedean ordered field). But difficulties arise as soon as one deals with transcendental functions: Thus to differentiate $\sin x$ à la Leibniz, one might wish to write:

$$\sin(x + dx) - \sin x = \sin x(\cos dx - 1) + \cos x \sin dx,$$

where dx is infinitesimal. But to write this assumes not only that sine has been defined for "numbers" of the form "real + infinitesimal," but also that this has been done in such a way that the addition law for the sine continues to hold. (In complex function theory analogous problems lead to analytic continuation and the principle of permanence of form.) It is this problem to which modern logic (more precisely, model theory) turns out to hold the key.

3. THE ROLE OF LOGIC

Leibniz postulated a system of numbers having the same *properties* as ordinary numbers but which included infinitesimals. So Leibniz had no problem with the differentiation of $\sin x$ as discussed above. Yet Leibniz' position seems absurd on its face. The ordinary real numbers obviously have at least one *property* not shared by Leibniz' desired extension. Namely, in the real numbers, there are no infinitesimals.

This paradox is avoided by specifying a formal language in the sense of modern logic (mercilessly precise in the same way that programming languages for computers are). Leibniz' principle is then reinterpreted: *there is an extension of the reals that includes infinitesimal elements and has the same properties as the real numbers insofar as those properties can be expressed in the specified formal language*. One concludes that the property of being infinitesimal cannot be so expressed, or, as we shall learn to say: *the set of infinitesimals is an external set*.

4. THREE TECHNIQUES

In nonstandard analysis one is working with two structures, the *standard universe* and the *nonstandard universe*. In addition there is a formal language that can be used to make assertions that are ambiguous in that they can refer to either of the two structures.

There are three main tools in nonstandard analysis. One is the *transfer principle*, which roughly states that the same assertions of the formal language are true in the standard universe as in the nonstandard universe. It is typically used by proving a desired result in the nonstandard universe, and then, noting that the result is expressible in the language, concluding that it holds in the standard universe as well.

Another technique is *concurrence*. This is a logical technique that guarantees that the extended structure contains all possible completions, compactifications, and so forth.

The third technique is *internality*. A set s of elements of the nonstandard universe is *internal* if s itself is an element of the nonstandard universe; otherwise, s is *external*. A surprisingly useful method of proof is one by reductio ad absurdum in which the contradiction is that some set one knows to be *external* would in fact be *internal* under the assumption being refuted.

Of course, the above discussion is only approximate. The reader should not expect these matters to be really clear until the detailed exposition has been presented.

5. MATHEMATICAL LOGIC AND RIGOR

Mathematicians who are perfectly content with the usual level of rigor in mathematical exposition in their own speciality and who are rarely troubled by philosophical qualms, sometimes exhibit symptoms of acute anxiety when the same standards are followed in applying mathematical logic. Hence this warning: Although logic is important in discussions of fundamental issues in the foundations of mathematics, these issues need not be injected when mathematical logic is being used just as another mathematical technique.

The use of logic in nonstandard analysis is in some ways analogous to the use of geometric language in mathematics. The intuition enabling us to read a "meaning" into a "sentence" of a formal language (technically speaking, a "sentence" is simply a finite sequence of objects called symbols) can be usefully compared to the intuition that enables us to perceive geometric truths by checking a diagram. Similar expository problems arise in the two cases. Just as writing detailed analytic proofs of things that are obvious from a diagram quickly degenerates into a tedious unrewarding task once one has learned how, so giving detailed proofs that specific sentences in a formal language "say" what they intuitively seem to say, quickly becomes superfluous. It is recommended that the reader ignore such proofs when they seem unnecessary and supply his own when they are omitted, if there is the least doubt about the matter.

6. NUMBERING OF THEOREMS

A reference to Theorem 2–8.1 is to Theorem 8.1 of Chapter 2, that is, to the first theorem of Section 8 of Chapter 2. When a reference is to a theorem in the same chapter, the chapter reference is omitted. Thus a reference to Theorem 7.2 within Chapter 2 is to Theorem 2–7.2.

1 UNIVERSES AND LANGUAGES

1. SETS AND RELATIONS

Our development of nonstandard analysis makes use of the convenient fact that the various objects and relations with which mathematics deals can all be construed as sets. Therefore, we begin with a brief survey of the notions of elementary set theory we will be using.

We write $x \in A$ to indicate that x is an element of the set A, $x \notin A$ to indicate the contrary. For sets A,B we write $A \subseteq B$ or $B \supseteq A$ to mean that $x \in A$ implies $x \in B$, and $A \subset B$ or $B \supset A$ to mean that $A \subseteq B$ but $A \neq B$.

We use the usual notations

$$\{x|---\}, \qquad \{x \in A|---\}$$

for defining particular sets. The finite set whose members are precisely $a_1, \ldots, a_n$ is written $\{a_1, \ldots, a_n\}$. The empty set, written $\varnothing$ is the only *set* that has no members. We also consider other objects, called *individuals*, which have no members, but then these cannot be sets. For any set A, we write

$$\mathscr{P}(A) = \{B|B \subseteq A\}.$$

$\mathscr{P}(A)$ is called the *power set* of A.

For any objects a,b we write:

$$\langle a,b\rangle = \{\{a\}, \{a,b\}\}.$$

$\langle a,b\rangle$ is called the *ordered pair* of a and b. Although this definition seems quite arbitrary, the name "ordered pair" is justified by the

LEMMA. $\langle a,b\rangle = \langle a',b'\rangle$ if and only if $a = a'$ and $b = b'$.

PROOF. We are given

$$\{\{a\}, \{a,b\}\} = \{\{a'\}, \{a',b'\}\}.$$

Equality of these sets means that they have the same members. So we are led to two cases:

CASE 1

$\{a\} = \{a'\}$, $\{a,b\} = \{a',b'\}$. In this case $a = a'$, so $\{a,b\} = \{a,b'\}$. If $a \neq b$, $\{a,b\}$ has two members. Hence so does $\{a,b'\}$ and $b = b'$. If $a = b$ then $a = b'$ so that again $b = b'$.

CASE 2

$\{a\} = \{a',b'\}$, $\{a,b\} = \{a'\}$. Then, $a = a' = b'$ and $a = b = a'$. So $a = a'$ and $b = a' = b'$. ■

We define the (*ordered*) *n-tuple* $\langle x_1,x_2,\ldots,x_n\rangle$, $n > 2$ by the recursion

$$\langle x_1,x_2,\ldots,x_n\rangle = \langle\langle x_1,x_2,\ldots,x_{n-1}\rangle, x_n\rangle.$$

It then follows by a trivial induction that $\langle x_1,\ldots,x_n\rangle = \langle y_1,\ldots,y_n\rangle$ if and only if $x_1 = y_1, x_2 = y_2, \ldots, x_n = y_n$. In order to include the case $n = 1$, we write $\langle x\rangle = x$. For any set X, we set

$$X^n = \{\langle x_1,\ldots,x_n\rangle | x_1,\ldots,x_n \in X\}.$$

If A and B are sets, we write

$$A \times B = \{\langle x,y\rangle | x \in A \text{ and } y \in B\}.$$

If $R \subseteq A \times B$, then R is called a *relation*. When $B = A$, we sometimes speak of a *relation on A*. For R a relation, we sometimes write xRy for $\langle x,y\rangle \in R$. By the domain of the relation R (written dom(R)), we mean the set of all x such that xRy for some y.

g is called a *function* (or a *mapping*) *from A into B* if $g \subseteq A \times B$ (g is a relation) and if for each $x \in A$ there is *exactly one* $y \in B$ such that $\langle x,y\rangle \in g$. For $x \in A$ we write $g(x)$ for the unique element of B with $\langle x,g(x)\rangle \in g$. For $C \subseteq A$, we set

$$g[C] = \{g(x) | x \in C\}.$$

$g[C]$ is called the *image* of C under g. Here $A = \text{dom}(g)$, the *domain* of g, and the set $g[A]$ is called the *range* of g. If $g[A] = B$, g is said to map A *onto B*. If $g(x) = g(y)$ implies $x = y$, then g is called *one-one*.

If f maps X^n into Y we call f an *n-ary function* (or *a function of n arguments*) *from X into Y*, and we write $f(x_1,\ldots,x_n)$ for $f(\langle x_1,\ldots,x_n\rangle)$.

Throughout, we write N for the set $\{0,1,2,3,\ldots\}$ of *natural numbers* or *nonnegative integers*. We write N^+ for the set $\{1,2,3,\ldots\}$ of positive integers.

Sometimes, when we are less interested in the domain of a function than its range, we shift our notation and language: we call the domain of the function an *index set*, and instead of a function we speak of an *indexed family*. Thus an indexed family with index set I is just a mapping X with domain I. In such a case, for each $i \in I$ we write X_i instead of $X(i)$ and we write $\{X_i | i \in I\}$ to indicate the family itself (i.e., the mapping). When the index set is clear from the context, we sometimes write simply $\{X_i\}$. If the index set is of the form $I = \{1,2, \ldots ,n\}$ or $\{0,1, \ldots ,n\}$, the family is called a *finite sequence*, and if the index set is N or N^+, the family is called an *infinite sequence*.

Let $\{X_i | i \in I\}$ be an indexed family of sets, that is, each X_i is a set. Then we write

$$\begin{aligned} \bigcup_{i \in I} X_i &= \bigcup\{X_i | i \in I\} \\ &= \{z | z \in X_i \text{ for some } i \in I\}, \\ \bigcap_{i \in I} X_i &= \bigcap\{X_i | i \in I\} \\ &= \{z | z \in X_i \text{ for all } i \in I\}, \\ \prod_{i \in I} X_i &= \prod\{X_i | i \in I\} \\ &= \{g | g \text{ is a function whose domain is } I \text{ such that } \\ &\qquad g(i) \in X_i \text{ for each } i \in I\}, \end{aligned}$$

and speak of the *union*, the *intersection*, and the *Cartesian product*, respectively, of the family. In the special case, $I = \{1,2, \ldots ,n\}$, we write

$$\begin{aligned} X_1 \cup X_2 \cup \cdots \cup X_n &= \bigcup_{i \in I} X_i, \\ X_1 \cap X_2 \cap \cdots \cap X_n &= \bigcap_{i \in I} X_i, \\ X_1 \times X_2 \times \cdots \times X_n &= \prod_{i \in I} X_i. \end{aligned}$$

If A and B are sets, we write:

$$A - B = \{x \in A | x \notin B\}.$$

Let $A_i \subseteq U$ for $i \in I$. Then one readily verifies the so-called *de Morgan* identities:

$$\begin{aligned} U - \bigcup_{i \in I} A_i &= \bigcap_{i \in I} (U - A_i), \\ U - \bigcap_{i \in I} A_i &= \bigcup_{i \in I} (U - A_i). \end{aligned}$$

2. FILTERS

We use some very simple properties of filters in the construction of our "nonstandard universe," and we now proceed to develop the necessary material.

DEFINITION. If I is any nonempty set, then $F \subseteq \mathscr{P}(I)$ is called a *filter on* I if

(1) $A \in F$ and $A \subseteq B$ implies $B \in F$,
(2) $A,B \in F$ implies $A \cap B \in F$, and
(3) $\varnothing \notin F$, $I \in F$.

Of course, (2) implies that the intersection of any finite number of elements of a filter is also in the filter.

EXAMPLE. Let $A_0 \subseteq I$, $A_0 \neq \varnothing$. Then as it is very easy to see,

$$F = \{A \in \mathscr{P}(I) | A_0 \subseteq A\}$$

is a filter on I.

DEFINITION. F is an *ultrafilter* on I if

(1) F is a filter on I.
(2) If $F \subseteq F_1$ and F_1 is a filter on I, then $F = F_1$.

EXAMPLE. For any $x \in I$, let $A_0 = \{x\}$. Then the family

$$F = \{A \in \mathscr{P}(I) | A_0 \subseteq A\}$$

is an ultrafilter. For, if $F \subseteq F_1$, and $A \in F_1 - F$, then $x \notin A$ (since otherwise $A \supseteq \{x\}$ and hence $A \in F$); thus $A \cap \{x\} = \varnothing \in F_1$, so that F_1 cannot be a filter.

The basic fact about the existence of ultrafilters is:

THEOREM 2.1. If F_0 is a filter on I, then there is an ultrafilter F on I such that $F \supseteq F_0$.

This theorem is, technically speaking, a weak form of the so-called axiom of choice. More precisely, it is known that:

(1) Theorem 2.1 cannot be proved from the axioms of set theory without the axiom of choice;
(2) it can be proved using the axiom of choice;
(3) it is weaker than the axiom of choice, that is, Theorem 2.1 does not imply the axiom of choice.

The proof given below uses Zorn's lemma, a convenient equivalent of the axiom of choice. The reader who chooses to ignore the proof and to simply use Theorem 2.1 itself as an axiom of set theory is of course welcome to do so. In fact this has some technical advantages: namely, a result whose classical proof uses the full axiom of choice (e.g., the existence of a non-Lebesgue measurable set of real numbers) can sometimes be seen to depend only on the weaker Theorem 2.1 when a nonstandard proof of it is given. However, we are not concerned with such matters here.

PROOF. Let $\mathscr{G} = \{G \supseteq F_0 | G \text{ is a filter on } I\}$. We wish to apply Zorn's Lemma to $\mathscr{G}$. Thus, let $\mathscr{D} \subseteq \mathscr{G}$ where $\mathscr{D}$ is linearly ordered under $\subseteq$ (i.e., $G_1,G_2 \in \mathscr{D}$ implies $G_1 \subseteq G_2$ or $G_2 \subseteq G_1$). Let $\tilde{G} = \bigcup_{G \in \mathscr{D}} G$. Then $\tilde{G} \supseteq F_0$, since for each $G \in \mathscr{D}$, $G \supseteq F_0$. Moreover, $\tilde{G}$ is a filter on I as we proceed to show:

(1) If $A \in \tilde{G}$ and $B \supseteq A$, then $A \in G$ for some $G \in \mathscr{D}$, where (since G is a filter) $B \in G$. Thus $B \in \tilde{G}$.
(2) If $A,B \in \tilde{G}$, then $A \in G_1$, $B \in G_2$, $G_1,G_2 \in \mathscr{D}$. Since $\mathscr{D}$ is linearly ordered, either $A,B \in G_1$ or $A,B \in G_2$; hence $A \cap B \in G_1$ or $A \cap B \in G_2$, so that $A \cap B \in \tilde{G}$.
(3) If $\varnothing \in \tilde{G}$, then $\varnothing \in G$ for some $G \in \mathscr{D}$, which is impossible; since $I \in G$ for all $G \in \mathscr{D}$, we have $I \in \tilde{G}$.

We have shown that $\mathscr{D}$ has an upper bound in $\mathscr{G}$. Hence by Zorn's lemma, $\mathscr{G}$ has a maximal element F. That is, $F \in \mathscr{G}$ and $F \subseteq F_1 \in \mathscr{G}$ implies $F = F_1$. Since $F \in \mathscr{G}$, we have $F \supseteq F_0$ and F is a filter. If $F \subseteq F_1$ where F_1 is a filter on I, then clearly $F_1 \in \mathscr{G}$, so that $F = F_1$. This completes the proof. ∎

DEFINITION. $G \subseteq \mathscr{P}(I)$ is called a *filter basis on I* if

(1) $\varnothing \notin G$,
(2) $A,B \in G$ implies $A \cap B \in G$,
(3) $G \neq \varnothing$.

THEOREM 2.2. If G is a filter basis on I, then there is a filter F on I such that $F \supseteq G$.

PROOF. Let $F = \{A \subseteq I | A \supseteq C \text{ for some } C \in G\}$. Clearly $G \subseteq F$. We now see that F is a filter.

(1) Let $A \in F$ and $B \supseteq A$. Then for some $C \in G$, $A \supseteq C$. Since $B \supseteq A$, we have $B \supseteq C$, and thus $B \in F$.
(2) Let $A,B \in F$. Then for $C_1,C_2 \in G$, $A \supseteq C_1$ and $B \supseteq C_2$, and, therefore, $A \cap B \supseteq C_1 \cap C_2$. Since G is a filter basis, $C_1 \cap C_2 \in G$ and hence $A \cap B \in F$.

(3) If $\varnothing \in F$, then $\varnothing \supseteq C$ for some $C \in G$, that is, $\varnothing \in G$ which would contradict G being a filter basis; thus $\varnothing \notin F$. Finally, since $G \neq \varnothing$, there is some $C_0 \in G$. Since $I \supseteq C_0$, we have $I \in F$. ■

Combining Theorem 2.1 and 2.2, we obtain the basic result on the existence of ultrafilters:

COROLLARY 2.3. If G is a filter basis on I, then there is an *ultrafilter* F on I such that $F \supseteq G$.

The usefulness of ultrafilters in our work lies in the fact that for each subset A of I either A or $I - A$ must lie in the ultrafilter. That is,

THEOREM 2.4. If F is an ultrafilter on I and $A \subseteq I$, then $A \in F$ or $I - A \in F$, but not both.

PROOF. Suppose $A \notin F$ and $I - A \notin F$. Let $G = \{X \subseteq I | A \cup X \in F\}$. We show that G is a filter, $G \supseteq F$, thereby obtaining a contradiction. To see that $F \subseteq G$, let $X \in F$. Since $X \subseteq A \cup X$ and F is a filter, $A \cup X \in F$. Thus, $X \in G$. We next verify that G is a filter on I.

(1) Let $X_1 \in G$ and $X_1 \subseteq X_2$. Then $A \cup X_1 \subseteq A \cup X_2$. Since $A \cup X_1 \in F$, we have $A \cup X_2 \in F$ and thus $X_2 \in G$.
(2) Let $X_1, X_2 \in G$. Then $A \cup X_1, A \cup X_2 \in F$. Since F is a filter $(A \cup X_1) \cap (A \cup X_2) \in F$. But $(A \cup X_1) \cap (A \cup X_2) = A \cup (X_1 \cap X_2)$ and hence $X_1 \cap X_2 \in G$.
(3) If $\varnothing \in G$ then $A \cup \varnothing = A \in F$. But we have assumed $A \notin F$; thus $\varnothing \notin G$. Since $A \cup I = I \in F$, we have $I \in G$.

Thus G is a filter and $G \supseteq F$. Since F is an ultrafilter, $G = F$. However,

$$A \cup (I - A) = I \in F.$$

Hence $I - A \in G$, and therefore, $I - A \in F$, a contradiction. Finally, if $A, I - A \in F$, then $\varnothing = A \cap (I - A) \in F$, which is impossible. ■

DEFINITION. Let F be an ultrafilter on I. Then we write

$$\mu_F(A) = \begin{cases} 1 & \text{if } A \in F \\ 0 & \text{if } A \notin F. \end{cases}$$

We call the map μ_F of $\mathscr{P}(I)$ into $\{0,1\}$, the *measure* induced by F. Where there is no ambiguity, we write μ for μ_F.

THEOREM 2.5. $\mu(\varnothing) = 0, \mu(I) = 1$.

PROOF. The proof is immediate from the definition of μ_F. ■

THEOREM 2.6. If $\mu(A_i) = 0, i = 1, \ldots, n$, then

$$\mu(A_1 \cup A_2 \cup \cdots \cup A_n) = 0.$$

PROOF. By Theorem 2.4,

$$I - A_i \in F \quad \text{for} \quad i = 1,2, \ldots, n.$$

Hence, letting $J = \{1,2, \ldots, n\}$

$$\bigcap_{i \in J} (I - A_i) \in F.$$

But by one of the de Morgan identities,

$$\bigcap_{i \in J} (I - A_i) = I - \bigcup_{i \in J} A_i.$$

Hence $\bigcup_{i \in J} A_i \notin F$, that is,

$$\mu(A_1 \cup \cdots \cup A_n) = 0. \quad \blacksquare$$

3. INDIVIDUALS AND SUPERSTRUCTURES

Applications of nonstandard analysis begin by specifying an appropriate set S of *individuals*. Typically S might be the set of points of a topological space, or the set of real numbers. It is technically useful to assume that the members of S *are not sets*; that is, if $x \in S$, then $x \neq \varnothing$ and x has no members (i.e., the assertion $t \in x$ is automatically false). Thus for $x \in S$, $t \subseteq x$ is always false and $\mathscr{P}(x) = \varnothing$. Henceforth, when we say, *let S be a set of individuals*, it should always be understood that the members of S are not sets.

Now, typically the members of S include objects that are defined to be sets in most currently popular mathematical expositions. For example, a real number might be defined to be a set of rational numbers, or the natural number $n+1$ might be identified (after von Neumann) with the set $\{0,1, \ldots, n\}$. How are we to reconcile this situation with our stipulation that the members of S are not to be sets? There are various ways to manage, of which we mention one:

We can assume that we have available some given sufficiently large set $\mathscr{I}$ of true individuals (sometimes called *urelemente*), about which we assume nothing except that they are not sets. Then we can simply embed S one-one in $\mathscr{I}$ and proceed to identify each element of S with its image in $\mathscr{I}$.

In any case, questions as to the true "nature" of mathematical objects—such as whether $\sqrt{2}$ is a set—are always irrelevant to mathematical practice. In mathematics one is exclusively concerned with abstract relationships rather than with the "internal constitution" of objects. It is this situation to which Bertrand Russell refers in his famous aphorism:

> *Mathematics is the subject in which we don't know*[1] *what we are talking about.*

Leaving philosophy and returning to mathematics, let S be a set of individuals. We proceed to show how to realize in a simple structure, all the sets (including relations and functions) that are needed in the usual mathematical constructions involving the elements of S.

First we define the hierarchy:

$$S_0 = S$$
$$S_{i+1} = S_i \cup \mathscr{P}(S_i),$$

so that S_i is defined for $i = 0,1,2,\ldots$. Then we define

$$\hat{S} = \bigcup_{i \in N} S_i = S_0 \cup S_1 \cup S_2 \cup \cdots.$$

$\hat{S}$ is then[2] called the *superstructure* with *individuals* S. Each element of S is called an *individual* of $\hat{S}$, and each element of $\hat{S} - S$ is called a set of $\hat{S}$. Note that $\varnothing \subseteq S$, so $\varnothing \in S_1$.

DEFINITION. Let $A \subseteq \hat{S}$. Then A is called *transitive in* $\hat{S}$ (or just *transitive* when no confusion will result) if for every $x \in A$, either $x \in S$ or $x \subseteq A$.

That A is transitive in $\hat{S}$ may be equivalently expressed in the form:

$$x \in A - S \quad \textit{and} \quad y \in x \quad \textit{implies} \quad y \in A.$$

This is because if A is a set, $x \subseteq A$ simply means that $y \in x$ implies $y \in A$.

LEMMA 1. Each S_i is transitive in $\hat{S}$.

PROOF. The proof is by induction. S_0 is obviously transitive. Suppose it is known that S_i is transitive, and let $x \in S_{i+1} - S$. Then either $x \in S_i - S$ or $x \in \mathscr{P}(S_i)$. In the first case $x \subseteq S_i$ by induction hypothesis, and in the second case $x \subseteq S_i$ by definition of power set. But $S_i \subseteq S_{i+1}$. Hence $x \subseteq S_{i+1}$. This shows that S_{i+1} is also transitive. ■

LEMMA 2. If $x \in y$ and $y \in S_i - S$, then $x \in S_{i-1}$.

PROOF. Since $y \notin S$, $i > 0$. By the definition of S_i, either $y \in S_{i-1}$ or $y \subseteq S_{i-1}$. But in the first case also, using Lemma 1, $y \subseteq S_{i-1}$. Thus, x is a member of y, a subset of S_{i-1} and, therefore, is a member of S_{i-1}. ■

[1] For our purposes, "don't care" would be more to the point.

[2] The notation $\hat{S}$ and the term *superstructure* are from Robinson and Zakon [16]. This use of the symbol ^ is totally unrelated to that in Machover–Hirschfeld [12].

LEMMA 3. If $x \in S_i - S$, then $\mathscr{P}(x) \in S_{i+2}$.

PROOF. By Lemma 1, $x \subseteq S_i$. Hence each subset of x is also a subset of S_i. That is, $\mathscr{P}(x) \subseteq \mathscr{P}(S_i) \subseteq S_{i+1}$. Hence $\mathscr{P}(x) \in S_{i+2}$.

LEMMA 4. $\hat{S}$ is transitive in $\hat{S}$.

PROOF. If $x \in \hat{S} - S$, then $x \in S_i - S$ for some i. The result follows from Lemma 1. ■

LEMMA 5. If $x \in \hat{S} - S$, then $\mathscr{P}(x) \in \hat{S}$.

PROOF. The proof is immediate from Lemma 3. ■

LEMMA 6. If $y \subseteq x$ and $x \in \hat{S} - S$, then $y \in \hat{S}$.

PROOF. The proof is immediate from Lemmas 4 and 5. ■

LEMMA 7. Let $x \in S_i - S$, and let $x \cap S = \varnothing$. (That is, let x be a set of sets.) Let $y = \bigcup_{z \in x} z$. Then $y \in S_i$.

PROOF. For any $t \in y$, we have for some z, $t \in z$ and $z \in x$. By Lemma 2, $z \in S_{i-1}$. By hypothesis, then $z \in S_{i-1} - S$. By Lemma 1, $t \in S_{i-1}$. We have thus shown that $y \subseteq S_{i-1}$. Hence $y \in S_i$. ■

LEMMA 8. Let $x \in \hat{S} - S$, and let $x \cap S = \varnothing$. Let $y = \bigcup_{z \in x} x$. Then $y \in \hat{S}$.

PROOF. The proof is immediate from Lemma 7. ■

LEMMA 9. $S_i \in S_{i+1}$. Therefore, each $S_i \in \hat{S}$.

PROOF. $S_i \subseteq S_i$; so $S_i \in \mathscr{P}(S_i)$. ■

LEMMA 10. If $a_1, \ldots, a_k \in S_i$, then $\{a_1, \ldots, a_k\} \in S_{i+1}$.

PROOF. $\{a_1, \ldots, a_k\}$ is a subset of S_i, hence a member of S_{i+1}. ■

LEMMA 11. If $a_1, \ldots, a_k \in \hat{S}$ then $\{a_1, \ldots, a_k\} \in \hat{S}$.

PROOF. Let $a_i \in S_{j_i}$, $i = 1,2 \ldots, k$. Let $N = \max(j_1, \ldots, j_k)$. Then each $a_i \in S_N$, and the result follows from Lemma 10. ■

LEMMA 12. If $x_1, \ldots, x_k \in \hat{S} - S$, then $x_1 \cup \cdots \cup x_k \in \hat{S}$.

PROOF. Let each $x_i \in S_N$, $i = 1,2, \ldots, k$. By Lemma 1, each $x_i \subseteq S_N$, and hence $(x_1 \cup \cdots \cup x_k) \subseteq S_N$. ■

LEMMA 13. If $x,y \in S_i$, then $\langle x,y \rangle \in S_{i+2}$.

PROOF. By Lemma 10, $\{x\}$, $\{x,y\} \in S_{i+1}$. Hence

$$\langle x,y \rangle = \{\{x\}, \{x,y\}\} \subseteq S_{i+1},$$

and $\langle x,y \rangle \in S_{i+2}$. ■

LEMMA 14. If $x_1, \ldots, x_n \in S_i$ and $n \geq 2$, then $\langle x_1, \ldots, x_n \rangle \in S_{i+2n-2}$.

PROOF. This is a trivial induction using Lemma 13. ■

THEOREM 3.1. If $X, Y \in \hat{S} - S$ then $X \times Y \in \hat{S}$. More precisely, if $X, Y \in S_i - S$, then $X \times Y \in S_{i+3}$. Also, $X^n \in \hat{S}$ for all $n \in N^+$.

PROOF. Let $X, Y \in S_i - S$. Then using Lemmas 1 and 13,

$$\begin{aligned} \langle x,y \rangle \in X \times Y \quad &\text{implies} \quad x \in X \text{ and } y \in Y \\ &\text{implies} \quad x,y \in S_i \\ &\text{implies} \quad \langle x,y \rangle \in S_{i+2}. \end{aligned}$$

Hence $X \times Y \subseteq S_{i+2}$ and $X \times Y \in S_{i+3}$. Similarly, $X^n \subseteq S_{i+2n-2}$; so $X^n \in S_{i+2n-1}$. ■

THEOREM 3.2. If $X, Y \in \hat{S} - S$ and f maps X into Y, then

(1) $f \in \hat{S}$;
(2) if $a \in X$, then $f(a) \in \hat{S}$;
(3) if $A \subseteq X$, then $f[A] \in \hat{S}$.

PROOF.

(1) Let $X, Y \in S_i - S$. Then, as in the proof of Theorem 3.1, $X \times Y \subseteq S_{i+2}$. Since $f \subseteq X \times Y$, $f \subseteq S_{i+2}$ and $f \in S_{i+3}$.
(2) Since S_i is transitive and $f(a) \in Y$, we have $f(a) \in S_i$.
(3) $f[A] \subseteq Y$ and $Y \subseteq S_i$. Hence $f[A] \subseteq S_i$ and $f[A] \in S_{i+1}$. ■

THEOREM 3.3. Let $J, V \in \hat{S} - S$, and for each $j \in J$, let $X_j \in V$. Then

(1) $\bigcup_{j \in J} X_j \in \hat{S}$,

(2) $\prod_{j \in J} X_j \in \hat{S}$.

PROOF.

(1) Let $J, V \in S_i$. Since $X_j \in V$ for each $j \in J$, and S_i is transitive, for each j we have $X_j \in S_i$, and so $X_j \subseteq S_i$. Hence $\left(\bigcup_{j \in J} X_j\right) \subseteq S_i$, that is, $\bigcup_{j \in J} X_j \in S_{i+1}$.
(2) Let $f \in \prod_{j \in J} X_j$. Then $f(j) \in X_j$ for each $j \in J$. Proceeding as in (1), each $f(j) \in S_i$. Since $J \subseteq S_i$, $f \subseteq S_i \times S_i$, and finally $f \in \mathscr{P}(S_i \times S_i)$. Thus

$$\prod_{j \in J} X_j \subseteq \mathscr{P}(S_i \times S_i),$$

and

$$\left(\prod_{j \in J} X_j\right) \in \mathscr{P}(\mathscr{P}(S_i \times S_i)).$$

Since $S_i \in \hat{S} - S$, the result follows from Lemma 5 and Theorem 3.1. ■

Let $r,s \in \hat{S}$. If there is one and only one $t \in \hat{S}$ for which

$$\langle s,t \rangle \in r,$$

we write $r \upharpoonright s = t$; if this is not the case (i.e., if there is no such t, or more than one), we set $r \upharpoonright s = \varnothing$. The operation $\upharpoonright$ is important because it possesses the properties:

(1) if r is a function and $s \in \operatorname{dom}(r)$, then $r \upharpoonright s = r(s)$;
(2) $r \upharpoonright s \in \hat{S}$ for all $r,s \in \hat{S}$.

4. UNIVERSES

Let S be a set of individuals.

DEFINITION. A subset of U of $\hat{S}$ is called a *universe with individuals* S if

(1) $\varnothing \in U$;
(2) $S \subseteq U$;
(3) if $x,y \in U$, then $\{x,y\} \in U$;
(4) U is transitive in $\hat{S}$.

The following is an obvious consequence of the previous section:

COROLLARY 4.1. $\hat{S}$ is a universe with individuals S.

It is important to note that a universe automatically satisfies the closure property:

THEOREM 4.2. If $r,s \in U$, then $\langle r,s \rangle \in U$ and $r \upharpoonright s \in U$.

PROOF. By definition, $\{r,s\}$ and $\{r\} = \{r,r\}$ are in U. Hence so is $\langle r,s \rangle = \{\{r\}, \{r,s\}\}$. If $r \upharpoonright s = \varnothing$, then, of course, $r \upharpoonright s \in U$. Otherwise, $r \quad s$ is the unique t for which $\langle s,t \rangle \in r$. Since U is transitive, $\langle s,t \rangle \in U$. Using the transitivity of U twice more, we get $\{s,t\} \in U$, $t \in U$. ■

The superstructure $\hat{S}$ is also called the *standard universe with individuals* S, or sometimes simply the *standard universe*. We show how to construct another universe, the so-called *nonstandard universe*, whose individuals *include* the elements of S and whose properties are closely related to those of $\hat{S}$.

For this purpose, let I be some nonempty index set (later, we shall want to specify I), let F be an ultrafilter on I, and let μ be the measure induced by F. We say that a property of elements of I holds a.e. (almost everywhere), or for almost all δ, if the set of elements of I for which the property holds, has measure 1 (or equivalently, that the set for which it fails has measure 0).

In our construction, we use functions f that map I into $\hat{S}$; for such f, we write $f_\delta = f(\delta)$ for each $\delta \in I$. For each $n \in N$, we let Z_n be the set of all functions f mapping I into $\hat{S}$ for which $f_\delta \in S_n$ a.e. Finally, let $Z = \bigcup_{n \in N} Z_n$.

There is a natural embedding of the standard universe $\hat{S}$ into Z. Namely, one *identifies* the element $r \in \hat{S}$ with the "constant" function such that

$$r_\delta = r \quad \text{for all} \quad \delta \in I.$$

For $f, g \in Z_0$, we write $f \sim g$ if $f_\delta = g_\delta$ a.e. We have

LEMMA 1. The relation $\sim$ is an equivalence relation on Z_0.

PROOF. It is clear that $f \sim f$ and that $f \sim g$ implies $g \sim f$. Suppose $f \sim g$ and $g \sim h$. Then the equations

$$f_\delta = g_\delta, \qquad g_\delta = h_\delta \tag{*}$$

hold almost everywhere. Now, the equation

$$f_\delta = h_\delta$$

holds whenever both of (*) hold; hence it fails at most on the union of the sets where each equation of (*) fails, that is, at most on the union of two sets of measure 0. Therefore, by Theorem 2.6, $f_\delta = h_\delta$ a.e., that is, $f \sim h$. ■

Theorem 2.6 is used in the proof of Lemma 1 to conclude that if each of a pair of conditions holds almost everywhere, so does the simultaneous assertion of both. We use Theorem 2.6 in this way repeatedly. (Of course, the same proof shows that the relation $f_\delta = g_\delta$ a.e. is an equivalence relation on all of Z, but we only need the result at this point for Z_0.)

For each $f \in Z_0$ we write

$$\bar{f} = \{g \in Z_0 | g \sim f\},$$

so that Z_0 is divided into the disjoint equivalence classes $\bar{f}$ by the relation $\sim$. We let

$$W = \{\bar{f} | f \in Z_0\}.$$

It is clear that for x,y in S (and hence via the embedding in Z_0) if $x \neq y$, then $\bar{x} \neq \bar{y}$. Hence we can identify each $x \in S$ with the corresponding element $\bar{x} \in W$, that is, we may now write $S \subseteq W$, and say that for $x \in S$ we have $\bar{x} = x$.

We are going to define a universe $\tilde{W}$ with individuals W, which we call the nonstandard universe (corresponding to $\hat{S}$). This involves forming the superstructure $\hat{W}$, which is of course defined by $W_0 = W$, $W_{i+1} = W_i \cup \mathscr{P}(W_i)$, $\hat{W} = \bigcup_{i \in N} W_i$. $\tilde{W}$ then consists of W together with certain of the sets of $\hat{W}$, to be specified later.

Some readers may still be disturbed by the fact that the elements of W, which, as we just saw, are "really" equivalence classes, are being regarded as nonsets in connection with the superstructure $\hat{W}$. They are invited to reread the beginning of Section 3 before proceeding.

In order to complete the definition of the nonstandard universe $\tilde{W}$, we must specify certain of the sets of $\hat{W}$ as belonging to $\tilde{W}$. In fact, we associate with each element $f \in Z_n$, a corresponding $\bar{f} \in W_n$, and these $\bar{f}$ constitute $\tilde{W}$.

$\bar{f}$ has already been defined for $f \in Z_0$, so that $\bar{f} \in W = W_0$. Let $i \geq 0$ and let us assume that $\bar{f}$ has already been defined for all $f \in Z_i$ in such a way that $\bar{f} \in W_i$. Then for $f \in Z_{i+1} - Z_i$ we define:

$\bar{f}$ is the set of all $\bar{g}$ such that $g \in Z_i$ and $g_\delta \in f_\delta$ a.e.

Then, by assumption, for each $\bar{g} \in \bar{f}$ we have $\bar{g} \in W_i$; hence $\bar{f} \subseteq W_i$ and $\bar{f} \in W_{i+1}$. We have thus given an inductive definition of $\bar{f}$ for all $f \in Z$; moreover $f \in Z_i$ implies $\bar{f} \in W_i$.

Finally, we define

$$\tilde{W} = \{\bar{f} | f \in Z\},$$

and call $\tilde{W}$ the *nonstandard universe* corresponding to $\hat{S}$. (Of course, $\tilde{W}$ also depends on the index set I and the ultrafilter F.) We shall see that $\tilde{W}$ is in fact a universe.

Since $\hat{S} \subseteq Z$ by our embedding, for each element $r \in \hat{S}$, there is a corresponding element $\bar{r} \in \tilde{W}$. Elements $\bar{r}$ for which $r \in \hat{S}$ are called the *standard elements* of $\tilde{W}$. The remaining elements of $\tilde{W}$ (if any) are called its *nonstandard elements*. In particular, the *standard individuals* are just the elements of S; the *nonstandard individuals* are those of $W - S$.

LEMMA 2. For $f,g \in Z$ we have $\bar{g} \in \bar{f}$ if and only if $g_\delta \in f_\delta$ a.e.

PROOF. If $\bar{g} \in \bar{f}$, then by definition $g_\delta \in f_\delta$ a.e. Conversely, let $f \in Z_n$, and let $g_\delta \in f_\delta$ a.e. Since f_δ is a set for almost all δ, it is impossible that $f_\delta \in S_0 = S$ a.e.; hence $n > 0$. Then $f_\delta \in S_n - S$ a.e. By Lemma 2 of Section 3 and Theorem 2.6, $g_\delta \in S_{n-1}$ a.e. Hence $g \in Z_{n-1}$. So by definition, $\bar{g} \in \bar{f}$. ■

LEMMA 3. If $f_\delta = g_\delta$ a.e. and $f \in Z_n$, then $g \in Z_n$.

PROOF. By Theorem 2.6, $f_\delta \in S_n$ a.e. and $f_\delta = g_\delta$ a.e. together imply that $g_\delta \in S_n$ a.e. ■

LEMMA 4. If $f,g \in Z$ and $f_\delta = g_\delta$ a.e., then $\bar{f} = \bar{g}$.

PROOF. Using Lemma 3, let $f,g \in Z_n$. If $n = 0$, the result follows by definition. So we can assume that $f,g \in Z_n - Z_0$, $n > 0$.

Since $f_\delta = g_\delta$ a.e., we have for any $k \in Z$ (by Theorem 2.6):

$$k_\delta \in f_\delta \text{ a.e. if and only if } k_\delta \in g_\delta \text{ a.e.}$$

Hence by Lemma 2,

$$\begin{aligned} \bar{k} \in \bar{f} \quad &\text{if and only if} \quad k_\delta \in f_\delta \text{ a.e.,} \\ &\text{if and only if} \quad k_\delta \in g_\delta \text{ a.e.,} \\ &\text{if and only if} \quad \bar{k} \in \bar{g}. \end{aligned}$$

Thus $\bar{f} = \bar{g}$. ■

LEMMA 5. If $f,g \in Z$ and $\bar{f} \subseteq \bar{g}$, then $f_\delta \subseteq g_\delta$ a.e.

PROOF. Since $\bar{f},\bar{g}$ are sets, $f,g \notin Z_0$. Let

$$A = \{\delta \in I \mid f_\delta \subseteq g_\delta\}.$$

We want to show that $\mu(A) = 1$. Suppose otherwise, that is, that $\mu(A) = 0$. We specify a function k as follows:

For $\delta \in A$, we set $k_\delta = \varnothing$; for $\delta \in I - A$, let k_δ be such that $k_\delta \in f_\delta$ but $k_\delta \notin g_\delta$. Since for some n, $f_\delta \in S_n$ a.e., we have (by Lemma 2 of Section 3 and Theorem 2.6) $k_\delta \in S_{n-1}$ a.e. Hence $k \in Z$. Since $k_\delta \in f_\delta$ a.e., we have by Lemma 2 that $\bar{k} \in \bar{f}$. By hypothesis $\bar{k} \in \bar{g}$, that is, $k_\delta \in g_\delta$ a.e. But $k_\delta \notin g_\delta$ for all $\delta \in I - A$, a set of measure 1. This contradiction proves the lemma. ■

Combining Lemmas 2, 4, and 5, we obtain

THEOREM 4.3. For $f,g \in Z$, we have

(1) $\bar{f} \in \bar{g}$ if and only if $f_\delta \in g_\delta$ a.e.,
(2) $\bar{f} = \bar{g}$ if and only if $f_\delta = g_\delta$ a.e.

PROOF. We need only check the "only if" part of (2). If $\bar{f} = \bar{g}$, then $\bar{f} \subseteq \bar{g}$ and $\bar{g} \subseteq \bar{f}$. Thus by Lemma 5, $f_\delta \subseteq g_\delta$ a.e. and $g_\delta \subseteq f_\delta$ a.e. By Theorem 2.6, $f_\delta = g_\delta$ a.e. ■

LEMMA 6. Let $f,g \in Z$. Let $k_\delta = \{f_\delta, g_\delta\}$ for each $\delta \in I$. Then $k \in Z$ and $\bar{k} = \{\bar{f},\bar{g}\}$.

PROOF. Let $f_\delta \in S_n$ a.e., $g_\delta \in S_n$ a.e. Then by Theorem 2.6 and Lemma 10 of Section 3, $k_\delta \in S_{n+1}$ a.e. Thus $k \in Z$.

Since $f_\delta \in k_\delta$ and $g_\delta \in k_\delta$ for all $\delta \in I$, it follows that $\bar{f},\bar{g} \in \bar{k}$. Let $\bar{h}$ be any element of $\bar{k}$. Then $h_\delta \in k_\delta$ a.e. Since $h_\delta \in k_\delta$ implies $h_\delta = f_\delta$ or $h_\delta = g_\delta$, it follows that either $h_\delta = f_\delta$ a.e. or $h_\delta = g_\delta$ a.e. That is, either $\bar{h} = \bar{f}$ or $\bar{h} = \bar{g}$. Hence $\bar{k} = \{\bar{f},\bar{g}\}$. ■

LEMMA 7. $\tilde{W}$ is transitive in $\hat{W}$.

PROOF. Let $\bar{f} \in \tilde{W}, \bar{f} \notin W$. Then by definition of $\bar{f}$, it consists of elements $\bar{g}$ of $\tilde{W}$, that is, $\bar{f} \subseteq \tilde{W}$. ■

THEOREM 4.4. $\tilde{W}$ is a universe with individuals W.

PROOF. Consulting the definition we see that the result is almost contained in Lemmas 6 and 7. What remains is to show that $\varnothing \in \tilde{W}$.

Now, $\varnothing \in S_1 \subseteq Z_1$. Then

$$\bar{\varnothing} = \{\bar{g} | g_\delta \in \varnothing \text{ a.e.}\} = \varnothing.$$

Hence $\varnothing \in \tilde{W}$. ■

LEMMA 8. Let $f,g,h \in Z$. Then $\bar{h} = \{\bar{f},\bar{g}\}$ if and only if $h_\delta = \{f_\delta,g_\delta\}$ a.e.

PROOF. Let $k_\delta = \{f_\delta,g_\delta\}$ for all δ. Thus by Lemma 6, $\bar{k} = \{\bar{f},\bar{g}\}$. Then by Theorem 4.3, $\bar{h} = \{\bar{f},\bar{g}\}$ if and only if $h_\delta = k_\delta$ a.e., which is the desired result.

LEMMA 9. Let $f,g,h \in Z$. Then $\bar{h} = \langle\bar{f},\bar{g}\rangle$ if and only if $h_\delta = \langle f_\delta,g_\delta\rangle$ a.e.

PROOF. Let $u_\delta = \{f_\delta\}$, $v_\delta = \{f_\delta,g_\delta\}$, $k_\delta = \{u_\delta,v_\delta\}$ for all $\delta \in I$. Then $k_\delta = \langle f_\delta,g_\delta\rangle$ for all $\delta \in I$. By Lemma 6, $\bar{u} = \{\bar{f}\}$, $\bar{v} = \{\bar{f},\bar{g}\}$, and $\bar{k} = \{\bar{u},\bar{v}\} = \langle\bar{f},\bar{g}\rangle$. Thus by Theorem 4.3, $\bar{h} = \langle\bar{f},\bar{g}\rangle$ if and only if $h_\delta = k_\delta$ a.e., the desired result. ■

LEMMA 10. Let $f,g \in Z$. Let $k_\delta = f_\delta \upharpoonright g_\delta$ for each $\delta \in I$. Then $k \in Z$, and $\bar{k} = \bar{f} \quad \bar{g}$.

PROOF. Let $f_\delta \in S_n$ a.e. For each $\delta \in I$, one of the following must occur:

(1) k_δ is the unique m such that $\langle g_\delta,m\rangle \in f_\delta$;
(2) $k_\delta = \varnothing$, and there exist l,m such that $l \neq m$ and $\langle g_\delta,l\rangle \in f_\delta$, $\langle g_\delta,m\rangle \in f_\delta$;
(3) $k_\delta = \varnothing$, and there is no m such that $\langle g_\delta,m\rangle \in f_\delta$.

Now one of (1), (2), (3) must hold a.e. (so that $k \in Z$), and we have three cases.

CASE 1

(1) *holds almost everywhere.* Then $\langle g_\delta,k_\delta\rangle \in f_\delta$ a.e. By Lemma 9, $\langle\bar{g},\bar{k}\rangle \in \bar{f}$. Suppose also that $\langle\bar{g},\bar{h}\rangle \in \bar{f}$. By Lemma 9 once again, $\langle g_\delta,h_\delta\rangle \in f_\delta$ a.e. Hence this last condition and (1) must hold *simultaneously* almost everywhere; that is, $h_\delta = k_\delta$ a.e. and $\bar{h} = \bar{k}$. This shows that $\bar{k} = \bar{f} \upharpoonright \bar{g}$.

CASE 2

(2) *holds almost everywhere.* Then, on a set of $\delta \in I$ of measure 1, we can define t_δ,u_δ such that $t_\delta \neq u_\delta$ and $\langle g_\delta,t_\delta\rangle \in f_\delta$ and $\langle g_\delta,u_\delta\rangle \in f_\delta$. On the set

of measure 0 on which (2) fails, we define $t_\delta = u_\delta = \varnothing$. Then $t_\delta, u_\delta \in S_n$ a.e., so that $t,u \in Z$. By Lemma 9, $\langle \bar{g},\bar{t} \rangle \in \bar{f}$, $\langle \bar{g},\bar{u} \rangle \in \bar{f}$, and by Theorem 4.3, $\bar{t} \neq \bar{u}$. Hence $\bar{f} \upharpoonright \bar{g} = \varnothing$. But $k_\delta = \varnothing$ a.e., so $\bar{k} = \varnothing$.

CASE 3

(3) *holds almost everywhere.* Suppose that $\langle \bar{g},\bar{h} \rangle \in \bar{f}$ for some $h \in Z$. Then $\langle g_\delta, h_\delta \rangle \in f_\delta$ a.e. But this is impossible because (3) holds almost everywhere. Thus $\bar{f} \upharpoonright \bar{g} = \varnothing = \bar{k}$. ■

LEMMA 11. Let $f,g,h \in Z$. Then $\bar{h} = \bar{f} \upharpoonright \bar{g}$ if and only if $h_\delta = f_\delta \upharpoonright g_\delta$ a.e.

PROOF. Let $k_\delta = f_\delta \upharpoonright g_\delta$ for all δ, so that by Lemma 10, $\bar{k} = \bar{f} \upharpoonright \bar{g}$. Theorem 4.3 now gives the desired result. ■

We summarize Lemmas 9 and 11 in

THEOREM 4.5. Let $f,g,h \in Z$. Then

(1) $\bar{h} = \langle \bar{f},\bar{g} \rangle$ if and only if $h_\delta = \langle f_\delta, g_\delta \rangle$ a.e.,
(2) $\bar{h} = \bar{f} \upharpoonright \bar{g}$ if and only if $h_\delta = f_\delta \upharpoonright g_\delta$ a.e.

Theorem 4.5 is a special case of the important Łos' theorem, and is used in proving it.

5. LANGUAGES

For each universe U, we construct a corresponding *language* $\mathscr{L}_U$ (or just $\mathscr{L}$) to be used in making assertions about U. Of course, these "languages" themselves are precisely defined mathematical entities.

Underlying each language $\mathscr{L}$ is a set $\mathscr{A}$, called the alphabet of $\mathscr{L}$, whose members are called *symbols.* (The reader should not draw philosophical conclusions from our use of the word "symbol." This usage no more represents a commitment as to the nature of the objects than, for example, the term "point" does in a "space"). We assume that no element of $\mathscr{A}$ is a finite sequence of other elements of $\mathscr{A}$; otherwise, $\mathscr{A}$ can consist of any objects whatever.

We write $\mathscr{A}$ as

$$\mathscr{A} = \mathscr{A}_1 \cup \mathscr{A}_2 \cup \mathscr{A}_3,$$

where $\mathscr{A}_1$ and $\mathscr{A}_2$ do not depend on the particular universe U. The sets $\mathscr{A}_1, \mathscr{A}_2$, and $\mathscr{A}_3$ are to be pairwise disjoint. The notation we use for our symbols is meant to suggest their intended "meaning." However, to emphasize the formal and abstract nature of our construction, symbols are printed in boldface.

(1) The symbols that belong to $\mathscr{A}_1$ are

$$\mathbf{=} \quad \mathbf{\in} \quad \mathbf{\neg} \quad \mathbf{\&} \quad \mathbf{\exists} \quad \mathbf{(} \quad \mathbf{)} \quad \mathbf{\langle} \quad \mathbf{\rangle} \quad \mathbf{\upharpoonright} \quad \mathbf{,}$$

(2) The symbols that belong to $\mathscr{A}_2$ called *variables* are the countably infinite set:

$$\mathbf{x}_1 \quad \mathbf{x}_2 \quad \mathbf{x}_3 \quad \cdots$$

(3) The set $\mathscr{A}_3$ is to be in one-one correspondence with the universe U. For each $b \in U$, the corresponding element in $\mathscr{A}_3$ is written **b** and called the *name* of b. The symbols in $\mathscr{A}_3$ are called *constants* (of $\mathscr{L}$). Of course, $\mathscr{A}_3$ is generally infinite and even uncountable.

A finite sequence of symbols of the alphabet of $\mathscr{L}$ is called an *expression* of $\mathscr{L}$. For example,

$$\mathbf{\&} \ \mathbf{=}$$

is an expression (of length 3).

An expression μ is called a *term* (of $\mathscr{L}$) if there is a finite sequence $\mu_1,\mu_2,\ldots,\mu_n = \mu$ of expressions such that for each i, $1 \leq i \leq n$ either

(1) μ_i is a variable, or
(2) μ_i is a constant (of $\mathscr{L}$), or
(3) $\mu_i = \langle \mu_j,\mu_k \rangle$ where $j,k < i$, or
(4) $\mu_i = (\mu_j \upharpoonright \mu_k)$ where $j,k < i$.

The expressions

$$(\mathbf{x}_1 \upharpoonright \mathbf{x}_2) \quad \text{and} \quad \langle (\mathbf{x}_3 \upharpoonright \mathbf{x}_2), \mathbf{x}_3$$

are examples of terms.

Note that in (3) and (4) we have written "$=$" and not "$\mathbf{=}$". For instance, (4) abbreviates the following: μ_i consists of "(" followed by the expression μ_j followed by "$\upharpoonright$" followed by the expression μ_k followed by ")". A term containing no variables is called a *closed* term.

An expression α is called a *formula* (of $\mathscr{L}$) if there is a finite sequence of expressions $\alpha_1,\alpha_2,\ldots,\alpha_n = \alpha$ such that for each i, $1 \leq i \leq n$, either

(1) $\alpha_i = (\mu \mathbf{=} \nu)$ where μ and ν are terms of $\mathscr{L}$, or
(2) $\alpha_i = (\mu \in \nu)$ where μ and ν are terms of $\mathscr{L}$, or
(3) $\alpha_i = \neg\alpha_j$ where $j < i$, or
(4) $\alpha_i = (\alpha_j \mathbin{\&} \alpha_k)$ where $j,k < i$, or
(5) $\alpha_i = (\exists \mathbf{x}_j \in \mu)\alpha_k$ where[1] $k < i$, and μ is a term of $\mathscr{L}$ in which $\mathbf{x}_j$ doesn't occur.

An example of a formula is

$$((\mathbf{x}_1 \mathbf{=} \mathbf{x}_2) \mathbin{\&} \neg(\mathbf{x}_1 \in \mathbf{x}_2)).$$

Expressions of the form $(\exists \mathbf{x}_i \in \mu)$ are called *existential quantifiers.*

[1] Usually μ will be a constant or a variable.

A given symbol may *occur* more than once in an expression. For example, the variable $\mathbf{x}_1$ occurs in the formula

$$((\mathbf{x}_1 \in \mathbf{x}_2) \,\&\, (\exists \mathbf{x}_1 \in \mathbf{b})(\mathbf{x}_1 \in \mathbf{x}_2))$$

as the third, tenth, and fifteenth symbol.

We speak of the first of these occurrences of $\mathbf{x}_1$ as *free* and of the other two as *bound* in the following sense:

An occurrence of a variable $\mathbf{x}_i$ in the formula α is called *bound* if there is a formula β that is part of α in which the given occurrence of $\mathbf{x}_i$ lies, where β is of the form $(\exists \mathbf{x}_i \in \mu)\gamma$. An occurrence of $\mathbf{x}_i$ in α that is not bound is called *free*.

A formula (of $\mathscr{L}$) in which no variable has a free occurrence is called a *sentence* (of $\mathscr{L}$).

For example, the formula

$$(\exists \mathbf{x}_1 \in \mathbf{b})\neg(\exists \mathbf{x}_2 \in \mathbf{c})(\mathbf{x}_1 \in \mathbf{x}_2)$$

is a sentence since all occurrences of its variables are bound.

Let μ be a term of $\mathscr{L}$, and let $\mathbf{x}_{i_1}, \ldots, \mathbf{x}_{i_k}$ include all of the variables that occur in μ (and possibly others as well). In such a case, we write

$$\mu = \mu(\mathbf{x}_{i_1}, \ldots, \mathbf{x}_{i_k}),$$

and then write $\mu(\mathbf{b}_1, \ldots, \mathbf{b}_k)$ for the *closed term* obtained by replacing each occurrence of $\mathbf{x}_{i_1}$ by $\mathbf{b}_1$, $\mathbf{x}_{i_2}$ by $\mathbf{b}_2$, and so forth. For α a formula of $\mathscr{L}$ we write

$$\alpha = \alpha(\mathbf{x}_{i_1}, \ldots, \mathbf{x}_{i_k})$$

when all variables which have *free* occurrences in α are included in the list $\mathbf{x}_{i_1}, \ldots, \mathbf{x}_{i_k}$. In this case, we write $\alpha(\mathbf{b}_1, \ldots, \mathbf{b}_k)$ for the *sentence* obtained by replacing each *free* occurrence of $\mathbf{x}_{i_1}$ by $\mathbf{b}_1$, $\mathbf{x}_{i_2}$ by $\mathbf{b}_2$, and so forth.

6. SEMANTICS

It is intended that each closed term of $\mathscr{L}_U$ "represents" a definite element of U, and that each sentence of $\mathscr{L}_U$ makes an assertion, true or false, about U. We proceed to make these notions precise, or as one says, to present the *semantics* of $\mathscr{L}_U$.

Let μ be a closed term of $\mathscr{L}_U$. We define a value $|\mu|_U$ (or just $|\mu|$) as follows:

(1) $|\mathbf{b}|_U = b$ for all $b \in U$,
(2) $|\langle \mu, \nu \rangle|_U = \langle |\mu|_U, |\nu|_U \rangle$,
(3) $|(\mu \upharpoonright \nu)|_U = (|\mu|_U) \upharpoonright (|\nu|_U)$.

This definition of $|\mu|_U$ is a recursion on the length of μ. That $|\mu|_U \in U$ follows at once (by induction) from Theorem 4.1.

Next we define recursively $U \models \alpha$ (read "α is *true* in U") where α is a *sentence* of $\mathscr{L}_U$

(1) $U \models (\mu = v)$ if and only if $|\mu|_U = |v|_U$,
(2) $U \models (\mu \in v)$ if and only if $(|\mu|_U) \in (|v|_U)$,
(3) $U \models \neg\alpha$ if and only if it is not the case that $U \models \alpha$,
(4) $U \models (\alpha \ \& \ \beta)$ if and only if $U \models \alpha$ and $U \models \beta$,
(5) $U \models (\exists \mathbf{x}_i \in \mu)\alpha(\mathbf{x}_i)$ if and only if $U \models \alpha(\mathbf{c})$ for some $c \in |\mu|_U$.

This definition is a recursion on the total number of occurrences in the sentence of the symbols: $\neg$, &, $\exists$. (1) and (2) give the definition of $U \models \alpha$ in case this number is 0, and (3), (4), and (5) reduce the determination of whether $U \models \alpha$, where this number >0, to that of $U \models \beta$ for one or more sentences β for which this number is smaller than for α. Note that the fact that we are dealing with *sentences* forces μ, v in (1) and (2), and μ in (5) to be *closed terms*, so that $|\mu|_U$, $|v|_U$ are defined.

A reader who has never seen such a "truth" definition before, may find it somewhat circular. And indeed, the definition is little help in deciding whether $U \models \alpha$ for complicated sentences α. (It is easy with a little practice to construct a sentence α such that $\hat{S} \models \alpha$ is equivalent to the Riemann hypothesis or to any other open question of classical mathematics.) What is important is that the intuitive notion of "truth" has been made precise in such a way that we can prove useful theorems about the notion.

When it is not the case that $U \models \alpha$, we write $U \not\models \alpha$ and say that α is *false* in U.

Formulas of $\mathscr{L}_U$ can be used not only to make assertions about U, but also to define subsets of U.

DEFINITION. Let $A \subseteq U$. Then A is called *definable* (or a *definable subset of* U) if there is a formula $\alpha = \alpha(\mathbf{x}_i)$ of $\mathscr{L}_U$ such that

$$A = \{b \in U \mid U \models \alpha(\mathbf{b})\}.$$

In this case, the formula α is called a *definition of A in $\mathscr{L}_U$*.

We now introduce by abbreviation certain basic logical operations not directly provided for in $\mathscr{L}_U$. Let α,β be arbitrary formulas of $\mathscr{L}_U$. Then we write

$$(\alpha \vee \beta) = \neg(\neg\alpha \ \& \ \neg\beta),$$
$$(\alpha \rightarrow \beta) = \neg(\alpha \ \& \ \neg\beta),$$
$$(\alpha \leftrightarrow \beta) = ((\alpha \rightarrow \beta) \ \& \ (\beta \rightarrow \alpha)).$$

Also, we write

$$(\forall \mathbf{x}_i \in \boldsymbol{\mu})\alpha = \neg(\exists \mathbf{x}_i \in \boldsymbol{\mu})\neg\alpha.$$

By a straightforward calculation, using the definition of $\models$, we obtain

$$\begin{aligned}
U \models (\alpha \vee \beta) &\quad \text{if and only if} \quad U \models \alpha \text{ or } U \models \beta \text{ (or both);} \\
U \models (\alpha \rightarrow \beta) &\quad \text{if and only if} \quad \text{either } U \not\models \alpha \text{ or } U \models \beta; \\
U \models (\alpha \leftrightarrow \beta) &\quad \text{if and only if} \quad \text{either } U \models \alpha \text{ and } U \models \beta, \\
&\qquad\qquad\qquad\quad \text{or } U \not\models \alpha \text{ and } U \not\models \beta; \\
U \models (\forall \mathbf{x}_i \in \boldsymbol{\mu})\alpha(\mathbf{x}_i) &\quad \text{if and only if} \quad U \models \alpha(\mathbf{c}) \text{ for all } c \in |\mu|_U.
\end{aligned}$$

7. ŁOS' THEOREM

In Section 6, we constructed a language $\mathscr{L}_U$ for each universe U. However, there are just two universes with which we are concerned: the standard universe $\hat{S}$ and the nonstandard universe $\tilde{W}$. Henceforth we write

$$\mathscr{L} = \mathscr{L}_{\hat{S}} \quad \textit{and} \quad {}^*\mathscr{L} = \mathscr{L}_{\tilde{W}}.$$

Also, if μ is a closed term of $\mathscr{L}$ we write $|\mu| = |\mu|_{\mathscr{L}}$, and if μ is a closed term of ${}^*\mathscr{L}$ we write $|\mu|_* = |\mu|_{{}^*\mathscr{L}}$. Finally, if α is a sentence of $\mathscr{L}$ we write $\models \alpha$ for $\hat{S} \models \alpha$, and if α is a sentence of ${}^*\mathscr{L}$ we write $* \models \alpha$ for $\tilde{W} \models \alpha$.

The expressive power of $\mathscr{L}$ is very great. In fact, as will become apparent, all of the propositions of classical mathematics are expressible in $\mathscr{L}$.

Let λ be a term (or formula) of $\mathscr{L}$. We let ${}^*\lambda$ be the term (or formula) of ${}^*\mathscr{L}$ obtained from λ by replacing each constant $\mathbf{b}$ in λ by the corresponding constant $\bar{\mathbf{b}}$. [We are making use of the fact that for each $b \in \hat{S}$, $\bar{b}$ is a (standard) element of $\tilde{W}$.] If, in particular, $b \in S$ for each b in λ, then ${}^*\lambda = \lambda$. [In this case, λ is simultaneously a term (or formula) of $\mathscr{L}$ and ${}^*\mathscr{L}$.]

What is crucial about our construction of the nonstandard universe $\tilde{W}$ is the close relation between the semantics of $\mathscr{L}$ and of ${}^*\mathscr{L}$ via the map just defined.

THEOREM 7.1. Let $\mu = \mu(\mathbf{x}_{i_1}, \ldots, \mathbf{x}_{i_n})$, $n \geq 0$, be a term of $\mathscr{L}$, let $g^1, g^2, \ldots, g^n \in Z$, and let

$$\bar{g} = |{}^*\mu(\bar{\mathbf{g}}^1, \ldots, \bar{\mathbf{g}}^n|_*.$$

Then

$$g_\delta = |\mu(\mathbf{g}_\delta^1, \ldots, \mathbf{g}_\delta^n)| \text{ a.e.}$$

PROOF. The proof is by induction on k the number of occurrences of "(" and "<" in μ.

First let $k = 0$. Then μ must be either a variable or a constant. If μ is a variable, then we can write

$$\mu = \mathbf{x}_{i_j} \text{ for some } j, 1 \leq j \leq n.$$

Then $\bar{g} = |\bar{\mathbf{g}}^j|_* = \bar{g}^j$. By Theorem 4.3,

$$g_\delta = g_\delta^j \text{ a.e.}$$

But for each $\delta \in I$,

$$g_\delta^j = |\mu(\mathbf{g}_\delta^1, \ldots, \mathbf{g}_\delta^n)|,$$

which gives the result. If μ is a constant, say $\mu = \mathbf{b}$, $b \in \hat{S}$, then $^*\mu = \bar{\mathbf{b}}$, and

$$\bar{g} = |\bar{\mathbf{b}}|_* = \bar{b}.$$

Thus by Theorem 4.3,

$$g_\delta = b \text{ a.e.}$$

(Of course, since $b \in \hat{S}$, $b_\delta = b$ for all δ.) But for all δ,

$$b = |\mathbf{b}| = |\mu(\mathbf{g}_\delta^1, \ldots, \mathbf{g}_\delta^n)|,$$

which again gives the result.

Having settled the case $k = 0$, we assume that $k > 0$ and consider two cases:

CASE 1

$\mu = \langle \nu,\eta \rangle$, where ν,η are terms having fewer than k occurrences of "(" and "$\langle$". Let

$$\bar{h} = |{}^*\nu(\bar{\mathbf{g}}^1,\bar{\mathbf{g}}^2, \ldots, \bar{\mathbf{g}}^n)|_*,$$
$$\bar{k} = |{}^*\eta(\bar{\mathbf{g}}^1,\bar{\mathbf{g}}^2, \ldots, \bar{\mathbf{g}}^n)|_*.$$

By the induction hypothesis,

$$h_\delta = |\nu(\mathbf{g}_\delta^1,\mathbf{g}_\delta^2, \ldots, \mathbf{g}_\delta^n)| \text{ a.e.}$$

and

$$k_\delta = |\eta(\mathbf{g}_\delta^1,\mathbf{g}_\delta^2, \ldots, \mathbf{g}_\delta^n)| \text{ a.e.}$$

Now, by the definition of $|\mu|_*$,

$$\bar{g} = \langle \bar{h},\bar{k} \rangle.$$

Hence by Theorem 4.5 and the definition of $|\mu|$,

$$\begin{aligned} g_\delta &= \langle h_\delta,k_\delta \rangle \text{ a.e.} \\ &= \langle |\nu(\mathbf{g}_\delta^1,\mathbf{g}_\delta^2, \ldots, \mathbf{g}_\delta^n)|, |\eta(\mathbf{g}_\delta^1,\mathbf{g}_\delta^2, \ldots, \mathbf{g}_\delta^n)| \rangle \text{ a.e.} \\ &= |\mu(\mathbf{g}_\delta^1,\mathbf{g}_\delta^2, \ldots, \mathbf{g}_\delta^n)| \text{ a.e.} \end{aligned}$$

CASE 2

$\mu \models (\nu \upharpoonright \eta)$, where ν,η are terms having fewer than k occurrences of "(" and "$\langle$". The proof in this case is entirely similar to the proof for Case 1. ■

COROLLARY 7.2. Let $\mu = \mu(\mathbf{x}_{i_1}, \ldots, \mathbf{x}_{i_n})$ be a term of $\mathscr{L}$, and let $g^1, \ldots, g^n \in Z$. For each $\delta \in I$, let

$$h_\delta = |\mu(\mathbf{g}_\delta^1, \ldots, \mathbf{g}_\delta^n)|.$$

Then $h \in Z$ and

$$\bar{h} = |{}^*\mu(\bar{\mathbf{g}}^1, \ldots, \bar{\mathbf{g}}^n)|_*.$$

PROOF. Assuming for the moment that $h \in Z$, let

$$\bar{g} = |{}^*\mu(\bar{\mathbf{g}}^1, \ldots, \bar{\mathbf{g}}^n)|_*.$$

By Theorem 7.1, $g_\delta = h_\delta$ a.e. so that by Theorem 4.3, $\bar{g} = \bar{h}$, the desired result.

We show that $h \in Z$ by an induction similar to that used in proving Theorem 7.1. If μ is a constant, $\mu = \mathbf{b}$, then $h_\delta = b$ for all δ and $h \in Z$; if $\mu = \mathbf{x}_{i_j}$, then $h = g^j \in Z$.

Proceeding inductively, let $\mu = \langle \nu, \eta \rangle$, and let

$$k_\delta = \nu(\mathbf{g}_\delta^1, \ldots, \mathbf{g}_\delta^n),$$
$$l_\delta = \eta(\mathbf{g}_\delta^1, \ldots, \mathbf{g}_\delta^n),$$

for all $\delta \in I$. By induction hypothesis we can assume $k_\delta, l_\delta \in S_m$ for some m and almost all δ. Then, by Section 3, Lemma 13, $h_\delta \in S_{m+2}$ for almost all δ.

Finally if $\mu = (\nu \upharpoonright \eta)$, the argument is almost the same except that $h_\delta \in S_m$ a.e. (as in the proof of Theorem 3.2). ■

THEOREM 7.3 (ŁOS' THEOREM). Let $\alpha = \alpha(x_{i_1}, \ldots, x_{i_n})$, $n \geq 0$ be a formula of $\mathscr{L}$, and let $g^1, g^2, \ldots, g^n \in Z$. Then

$$* \models {}^*\alpha(\bar{\mathbf{g}}^1, \ldots, \bar{\mathbf{g}}^n)$$

if and only if

$$\models \alpha(\mathbf{g}_\delta^1, \ldots, \mathbf{g}_\delta^n) \text{ a.e.}$$

PROOF. The proof is by induction on the number k of occurrences in α of the symbols $\neg$, &, $\exists$. If this number is 0, then

$$\alpha = (\mu = \nu) \quad \text{or} \quad \alpha = (\mu \in \nu).$$

for suitable terms μ, ν. First assume that $\alpha = (\mu = \nu)$, so that ${}^*\alpha = ({}^*\mu = {}^*\nu)$. Then using Theorems 7.1 and 4.3,

$$* \models {}^*\alpha(\bar{\mathbf{g}}^1, \ldots, \bar{\mathbf{g}}^n)$$

if and only if

$$|{}^*\mu(\bar{\mathbf{g}}^1, \ldots, \bar{\mathbf{g}}^n)|_* = |{}^*\nu(\bar{\mathbf{g}}^1, \ldots, \bar{\mathbf{g}}^n)|_*,$$

if and only if

$$|\mu(\mathbf{g}_\delta^1, \ldots, \mathbf{g}_\delta^n)| = |\nu(\mathbf{g}_\delta^1, \ldots, \mathbf{g}_\delta^n)| \text{ a.e.},$$

[1] For a discussion of Łos' Theorem in a more general context, see Bell and Slomson [1].

if and only if

$$\models \alpha(\mathbf{g}^1_\delta, \ldots, \mathbf{g}^n_\delta) \text{ a.e.}$$

If $\alpha = (\boldsymbol{\mu} \in \mathbf{v})$, the proof is the same, replacing the two "=" by "∈".

Assuming $k > 0$, there are three cases to consider:

CASE 1

$\alpha = \neg\beta$. Then, using the induction hypothesis,

$$* \models {}^*\alpha(\overline{\mathbf{g}}^1, \ldots, \overline{\mathbf{g}}^n)$$

if and only if

$$* \not\models \beta(\overline{\mathbf{g}}^1, \ldots, \overline{\mathbf{g}}^n),$$

if and only if it is not the case that

$$\models \beta(\mathbf{g}^1_\delta, \ldots, \mathbf{g}^n_\delta) \text{ a.e.,}$$

if and only if

$$\not\models \beta(\mathbf{g}^1_\delta, \ldots, \mathbf{g}^n_\delta) \text{ a.e.,}$$

if and only if

$$\models \alpha(\mathbf{g}^1_\delta, \ldots, \mathbf{g}^n_\delta) \text{ a.e.}$$

CASE 2

$\alpha = (\beta \mathbin{\&} \gamma)$. Then, using the induction hypothesis,

$$* \models {}^*\alpha(\overline{\mathbf{g}}^1, \ldots, \overline{\mathbf{g}}^n)$$

if and only if

$$* \models {}^*\beta(\overline{\mathbf{g}}^1, \ldots, \overline{\mathbf{g}}^n) \quad \text{and} \quad * \models {}^*\gamma(\overline{\mathbf{g}}^1, \ldots, \overline{\mathbf{g}}^n),$$

if and only if

$$\models \beta(\mathbf{g}^1_\delta, \ldots, \mathbf{g}^n_\delta) \text{ a.e.} \quad \text{and} \quad \models \gamma(\mathbf{g}^1_\delta, \ldots, \mathbf{g}^n_\delta) \text{ a.e.,}$$

if and only if

$$\models \alpha(\mathbf{g}^1_\delta, \ldots, \mathbf{g}^n_\delta) \text{ a.e.}$$

CASE 3

$\alpha(\mathbf{x}_{i_1}, \ldots, \mathbf{x}_{i_n}) = (\exists \mathbf{x}_k \in \mu)\beta(\mathbf{x}_k, \mathbf{x}_{i_1}, \ldots, \mathbf{x}_{i_n})$, where $\mu = \mu(x_{i_1}, \ldots, x_{i_n})$ is a term of $\mathscr{L}$. First suppose that $* \models {}^*\alpha(\overline{\mathbf{g}}^1, \ldots, \overline{\mathbf{g}}^n)$, and let $\overline{h} = |{}^*\mu(\overline{\mathbf{g}}^1, \ldots, \overline{\mathbf{g}}^n)|_*$. Then by the semantics of ${}^*\mathscr{L}$, there is a $\overline{g} \in \widetilde{W}$ such that

$$\overline{g} \in \overline{h} \quad \text{and} \quad * \models {}^*\beta(\overline{\mathbf{g}}, \overline{\mathbf{g}}^1, \ldots, \overline{\mathbf{g}}^n).$$

Then by Theorem 4.3 and the induction hypothesis,

$$g_\delta \in h_\delta \text{ a.e.} \quad \text{and} \quad \models \beta(\mathbf{g}_\delta, \mathbf{g}^1_\delta, \ldots, \mathbf{g}^n_\delta) \text{ a.e.}$$

Also, by Theorem 7.1,

$$h_\delta = |\mu(g_\delta^1, \ldots, g_\delta^n)| \text{ a.e.}$$

Thus for almost all $\delta \in I$, we have

$$\vDash \alpha(\mathbf{g}_\delta^1, \ldots, \mathbf{g}_\delta^n),$$

the desired result.

Conversely, let

$$\vDash \alpha(\mathbf{g}_\delta^1, \ldots, \mathbf{g}_\delta^n) \text{ a.e.,}$$

That is, let this condition hold for all $\delta \in A$ where $\mu(A) = 1$. Let $h_\delta = |\mu(\mathbf{g}_\delta^1, \ldots, \mathbf{g}_\delta^n)|$ for all $\delta \in I$. By Corollary 7.2, $h \in Z$, say $h_\delta \in S_m$ a.e. Then for each $\delta \in A$, there is a $t = t(\delta)$ such that $t \in h_\delta$, and

$$\vDash \beta(\mathbf{t}, \mathbf{g}_\delta^1, \ldots, \mathbf{g}_\delta^n).$$

Let g mapping I into $\hat{S}$ be defined by $g_\delta = t(\delta)$ for $\delta \in A$, $g_\delta = \varnothing$ for $\delta \notin A$. Then by the transitivity of S_m, $g_\delta \in S_m$ for each $\delta \in A$ for which $h_\delta \in S_m$; hence $g_\delta \in S_m$ a.e., so that $g \in Z$, and we have for almost all δ:

$$g_\delta \in h_\delta \quad \text{and} \quad \vDash \beta(\mathbf{g}_\delta, \mathbf{g}_\delta^1, \ldots, \mathbf{g}_\delta^n).$$

By Theorem 4.3 and the induction hypothesis,

$$\bar{g} \in \bar{h} \quad \text{and} \quad * \vDash {}^*\beta(\bar{\mathbf{g}}, \bar{\mathbf{g}}^1, \ldots, \bar{\mathbf{g}}^n).$$

Also, by Corollary 7.2, $\bar{h} = |{}^*\mu(\bar{\mathbf{g}}^1, \ldots, \bar{\mathbf{g}}^n)|_*$. Hence

$$* \vDash {}^*\alpha(\bar{\mathbf{g}}^1, \ldots, \bar{\mathbf{g}}^n). \quad \blacksquare$$

The case $n = 0$ of Łos' theorem is the important:

TRANSFER PRINCIPLE. *Let α be a sentence of $\mathscr{L}$. Then*

$$* \vDash {}^*\alpha \quad \textit{if and only if} \quad \vDash \alpha.$$

The Transfer Principle provides one of the basic tools of nonstandard analysis. A mathematical theorem that is equivalent to $\vDash \alpha$ for some sentence α of $\mathscr{L}$ can be proved by showing instead that $* \vDash {}^*\alpha$. Note that if α only contains constants $\mathbf{b}$ for $b \in S$, so that ${}^*\alpha = \alpha$, the Transfer Principle assumes the simpler form:

$$* \vDash \alpha \quad \textit{if and only if} \quad \vDash \alpha.$$

Łos' Theorem can be used to enable us to define an important operation on definable subsets of $\hat{S}$.

THEOREM 7.4. Let $\alpha = \alpha(\mathbf{x}_1)$, $\beta = \beta(\mathbf{x}_1)$ be formulas of $\mathscr{L}$ where

$$\{b \in \hat{S} | \vDash \alpha(\mathbf{b})\} = \{b \in \hat{S} \vDash \beta(\mathbf{b})\}.$$

Then,

$$\{\bar{g} \in \tilde{W} | * \vDash {}^*\alpha(\bar{\mathbf{g}})\} = \{\bar{g} \in \tilde{W} | * \vDash {}^*\beta(\bar{\mathbf{g}})\}.$$

PROOF. We have, using Łos' Theorem,

$$\begin{aligned} * \vDash {}^*\alpha(\bar{\mathbf{g}}) \quad &\text{if and only if} \quad \vDash \alpha(\mathbf{g}_\delta) \text{ a.e.,} \\ &\text{if and only if} \quad \vDash \beta(\mathbf{g}_\delta) \text{ a.e.,} \\ &\text{if and only if} \quad * \vDash {}^*\beta(\bar{\mathbf{g}}). \quad \blacksquare \end{aligned}$$

This theorem justifies the

DEFINITION. Let $A = \{b \in \hat{S} | \vDash \alpha(\mathbf{b})\}$ where α is a formula of $\mathscr{L}$. Then we write

$$^*A = \{b \in \tilde{W} | * \vDash {}^*\alpha(\mathbf{b})\}.$$

The theorem just proved guarantees that $^*A$ depends only on the set A and not on the particular formula α used to define it.

COROLLARY 7.5. Let r be a set of $\hat{S}$. Then r is a definable subset of $\hat{S}$ and $^*r = \bar{r}$.

PROOF. We have

$$r = \{b \in \hat{S} | \vDash (\mathbf{b} \in \mathbf{r})\}.$$

Hence

$$\begin{aligned} ^*r &= \{\bar{g} \in \tilde{W} | * \vDash (\bar{\mathbf{g}} \in \bar{\mathbf{r}})\} \\ &= \{\bar{g} \in \tilde{W} | \bar{g} \in \bar{r}\} = \bar{r}. \quad \blacksquare \end{aligned}$$

On the other hand, $\hat{S}$ itself is a definable subset of $\hat{S}$, using, for example, $\alpha(\mathbf{x}_1) = (\mathbf{x}_1 = \mathbf{x}_1)$,

$$\hat{S} = \{b \in \hat{S} | \vDash (\mathbf{b} = \mathbf{b})\},$$

(although $\hat{S}$ is certainly not[1] an element of $\hat{S}$). Thus we have

$$^*(\hat{S}) = \{b \in \tilde{W} | * \vDash (\mathbf{b} = \mathbf{b})\} = \tilde{W}.$$

From now on we write $U = \hat{S}$, that is, *we use the letter U exclusively to represent the standard universe.* Then $^*U = \tilde{W}$, and we have available the pleasant notation $^*U$ for the nonstandard universe.

Again, for each $i \geq 0$, $U - S_i = \hat{S} - S_i \notin \hat{S}$. (If $\hat{S} - S_i \in \hat{S}$, then since $S_i \in \hat{S}$, we would have $\hat{S} = ((\hat{S} - S_i) \cup S_i) \in \hat{S}$.)

$U - S_i$ is definable using $\neg(\mathbf{x}_1 \in \mathbf{S}_i)$:

$$U - S_i = \{b \in U | \vDash \neg(\mathbf{x}_1 \in \mathbf{S}_i)\}.$$

[1] The usual axioms of set theory forbid $x \in x$. However even without such a prohibition it is clear that $\hat{S} \notin \hat{S}$: Namely, if $\hat{S} \in \hat{S}$, then $\hat{S} \in S_n$ for some n. Then $S_{n+1} \in S_n$ and even $S_{n+1} \subseteq S_n$. But $\mathscr{P}(S_n) \subseteq S_{n+1}$, and this implies that S_{n+1} has a larger cardinality than S_n.

Then

$$*(U - S_i) = \{b \in *U | * \models \neg(\mathbf{x}_1 \in \bar{\mathbf{S}}_i)\}$$
$$= *U - \bar{S}_i = *U - *S_i,$$

using Corollary 7.5.

We also observe that by Theorem 4.3, $\bar{f} \in \bar{S}$ if and only if $f_\delta \in S$ a.e., that is, if and only if $f \in Z_0$. But by the definition of W, $\bar{f} \in W$ if and only if $f \in Z_0$. Hence $\bar{S} = *S = W$, and by the case $i = 0$ of the preceding paragraph,

$$*(U - S) = *U - W.$$

Remark on star (*) *versus bar* (¯): Since for elements x of $U - S$, $\bar{x} = *x$, we may use whichever of the two notations we please. In fact, the bar mapping of Z into $\tilde{W}$ is important only in the construction of $\tilde{W}$ and is not used in the applications. Hence we use the * notation almost exclusively from now on. For the sake of completeness, we also define $*x = x$ for $x \in S$. The *standard elements* of $*U$ are then just those of the form $*x$ for $x \in U$.

THEOREM 7.6. If $A \subseteq S$, then $A \subseteq *A$ and $*A \cap S = A$.

PROOF. Let $a \in A$. Then $\models(\mathbf{a} \in \mathbf{A})$. By the Transfer Principle, $* \models (\mathbf{a} \in *\mathbf{A})$ since $*a = \bar{a} = a$. Therefore, $a \in *A$. This shows that $A \subseteq *A$. Hence $A \subseteq *A \cap S$.

Conversely, let $a \in *A \cap S$. Since $a \in S$, $*a = a$. Then $* \models (*\mathbf{a} \in *\mathbf{A})$, so that by the Transfer Principle, $\models \mathbf{a} \in \mathbf{A}$, that is, $a \in A$. ■

Using Corollary 7.5, the Theorems 4.3 and 4.5 yield at once:

THEOREM 7.7. Let $x,y \in U$. Then

(1) $x = y$ if and only if $*x = *y$,
(2) $x \in y$ if and only if $*x \in *y$,
(3) $*\langle x,y\rangle = \langle *x,*y\rangle$,
(4) $*(x \upharpoonright y) = (*x \upharpoonright *y)$.

THEOREM 7.8. Let A,B be definable subsets of U. Then

$$(A \cup B)^* = *A \cup *B,$$
$$(A \cap B)^* = *A \cap *B,$$
$$(A - B)^* = *A - *B.$$

PROOF. Let $A = \{b \in U | \models \alpha(\mathbf{b})\}$, $B = \{b \in U | \models \beta(\mathbf{b})\}$. Then

$$A \cup B = \{b \in U | \models (\alpha(\mathbf{b}) \vee \beta(\mathbf{b}))\},$$
$$A \cap B = \{b \in U | \models (\alpha(\mathbf{b}) \,\&\, \beta(\mathbf{b}))\},$$
$$A - B = \{b \in U | \models (\alpha(\mathbf{b}) \,\&\, \neg\beta(\mathbf{b}))\}.$$

Hence

$$
\begin{aligned}
{}^*(A \cup B) &= \{b \in {}^*U | {}^* \models ({}^*\alpha(\mathbf{b}) \vee {}^*\beta(\mathbf{b}))\} \\
&= \{b \in {}^*U | {}^* \models {}^*\alpha(\mathbf{b})\} \cup \{b \in {}^*U | {}^* \models {}^*\beta(\mathbf{b})\} \\
&= {}^*A \cup {}^*B.
\end{aligned}
$$

The remaining cases are similar. ■

COROLLARY 7.9. ${}^*\varnothing = \varnothing$, ${}^*\{a_1, \ldots, a_k\} = \{{}^*a_1, \ldots, {}^*a_k\}$.

PROOF. Using the theorem,

$$
{}^*\varnothing = {}^*(U - U) = {}^*U - {}^*U = \varnothing.
$$

Also, if $c = \{a\} = \{b \in U | \models (\mathbf{b} = \mathbf{a})\}$, then ${}^*c = \{b \in U | {}^* \models (\mathbf{b} = {}^*\mathbf{a})\} = \{{}^*a\}$. This is the case $k = 1$ of what we wish to show. Proceeding by induction and using the theorem,

$$
\begin{aligned}
{}^*\{a_1, \ldots, a_k, a_{k+1}\} &= {}^*(\{a_1, \ldots, a_k\} \cup \{a_{k+1}\}) \\
&= {}^*\{a_1, \ldots, a_k\} \cup {}^*\{a_{k+1}\} \\
&= \{{}^*a_1, \ldots, {}^*a_k\} \cup \{{}^*a_{k+1}\} \\
&= \{{}^*a_1, \ldots, {}^*a_k, {}^*a_{k+1}\}. \quad ■
\end{aligned}
$$

THEOREM 7.10. ${}^*U = \bigcup_{i=0}^{\infty} {}^*(S_i)$.

PROOF. Since each $S_i \in U$, we have ${}^*S_i \in {}^*U$. And since ${}^*U$ is transitive, ${}^*S_i \subseteq {}^*U$. Thus $\bigcup_{i=0}^{\infty} {}^*S_i \subseteq {}^*U$.

Conversely, let $\bar{f} \in {}^*U = \tilde{W}$. Then $f \in Z_i$ for some $i \in N$. Since $f_\delta \in S_i$ a.e., Theorem 4.3 yields: $\bar{f} \in \bar{S}_i = {}^*S_i$. ■

We now introduce conventions that make formulas of $\mathscr{L}$ and ${}^*\mathscr{L}$ shorter and easier to read.

(1) We freely use any boldface letters as variables; this enables us to reflect ordinary mathematical usage (e.g., **n** for a natural number, **f** for a function, etc.).
(2) Parentheses are freely omitted in formulas where they are not needed for comprehension.
(3) We use English phrases and mathematical notation in combination, printed in boldface and containing appropriate variables, to stand for corresponding formulas of $\mathscr{L}$, in a manner that will be made clear by the examples below.

1. **X is a set**. This simply stands for the formula

$$
\alpha = \alpha(\mathbf{X}) = (\mathbf{X} = \varnothing) \vee (\exists \mathbf{x} \in \mathbf{X})(\mathbf{x} = \mathbf{x}).
$$

Of course, A is in fact a set (of S) just in case $\models \alpha(\mathbf{A})$, that is, just in case

$$\models \mathbf{A \text{ is a set}}.$$

2. $\mathbf{X \subseteq Y}$. The corresponding formula (containing two free variables) is

$$\mathbf{X \text{ is a set} \;\&\; Y \text{ is a set} \;\&\; (\forall x \in X)(x \in Y).}$$

Naturally, $\models(\mathbf{A \subseteq B})$ just in case A,B are sets and $A \subseteq B$.

3. $\mathbf{Z = X \times Y}$ stands for the formula:

$$\mathbf{(\forall z \in Z)(\exists x \in X)(\exists y \in Y)(z = \langle x,y\rangle) \;\&\; (\forall x \in X)(\forall y \in Y)(\quad z \in Z)(z = \langle x\ y\rangle).}$$

4. **f maps X into Y** abbreviates

$$\mathbf{X \text{ is a set} \;\&\; Y \text{ is a set} \;\&\; f \text{ is a set} \;\&\; (\forall t \in f)(\exists x \in X)(\exists y \in Y)(t = \langle x\ y\rangle)}$$
$$\mathbf{\&\; (\forall x \in X)(\forall y \in Y)(\langle x,y\rangle \in f \leftrightarrow y = f \quad x).}$$

f maps X onto Y abbreviates

$$\mathbf{f \text{ maps } X \text{ into } Y \;\&\; (\forall y \in Y)(\exists x \in X)(y = f \restriction x).}$$

In applications it is often necessary to compute $*A$, $*\alpha$ for various specific sets A and formulas α. Let us consider the above examples in this connection:

1. For the formula **X is a set**, we have $*\alpha = \alpha$ (since $*\varnothing = \varnothing$). In the standard universe α defines the set $U - S$. Hence in the nonstandard universe it defines $*(U - S)$, which as we have seen is just $*U - W$.

2. Suppose $A \subseteq B$, $A,B \in U$. Then, $\models \mathbf{A \subseteq B}$. By the Transfer Principle

$$* \models \mathbf{*A \text{ is a set} \;\&\; *B \text{ is a set} \;\&\; (\forall x \in *A)(x \in *B).}$$

Using the semantics of $*\mathscr{L}$, we conclude that $*A \subseteq *B$. We have proved

THEOREM 7.11. If $A,B \in U$ and $A \subseteq B$, then $*A \subseteq *B$.

THEOREM 7.12. Let $B \in U$, $A \in *U$, and let $A \subseteq *B$. Then $A \in *\mathscr{P}(B)$.

PROOF. By Theorem 7.10, $A \in *S_i$ for some $i \in N$. Let $P = \mathscr{P}(B)$. We have

$$\models \mathbf{(\forall X \in S_i)(X \subseteq B \to X \in P)}$$

By the Transfer Principle,

$$* \models \mathbf{(\forall X \in *S_i)(X \subseteq *B \to X \in *P).}$$

The semantics of $*\mathscr{L}$ yields $A \in *P$. ■

3. Since the formula abbreviated by $\mathbf{Z = X \times Y}$ contains no constants, the * operation leaves it unchanged. Using the Transfer Principle we conclude that

$$(A \times B)^* = *A \times *B.$$

4. Let f map A *onto* B. Then using the Transfer Principle we have

$$* \models (\forall \mathbf{t} \in {}^*\mathbf{f})(\exists \mathbf{x} \in {}^*\mathbf{A})(\exists \mathbf{y} \in {}^*\mathbf{B})\mathbf{t} = \langle \mathbf{x},\mathbf{y} \rangle,$$
$$* \models (\forall \mathbf{x} \in {}^*\mathbf{A})(\forall \mathbf{y} \in {}^*\mathbf{B})(\langle \mathbf{x},\mathbf{y} \rangle \in {}^*\mathbf{f} \leftrightarrow \mathbf{y} = {}^*\mathbf{f} \upharpoonright \mathbf{x}),$$
$$* \models (\forall \mathbf{y} \in {}^*\mathbf{B})(\exists \mathbf{x} \in {}^*\mathbf{A})\mathbf{y} = {}^*\mathbf{f} \upharpoonright \mathbf{x}.$$

We conclude that *$^*f$ is a function that maps $^*A$ onto $^*B$*. The fact that functions are mapped into functions by the * operation is quite important. In fact, we have

THEOREM 7.13. Let $f \in U$ be a function and let $C \subseteq \mathrm{dom}(f)$. Then

$$^*(f[C]) = {}^*f[{}^*C].$$

PROOF. We have

$$f[C] = \{b \in U | \models (\exists \mathbf{c} \in \mathbf{C})(\mathbf{b} = \mathbf{f} \upharpoonright \mathbf{c})\}.$$

Hence

$$\begin{aligned} ^*(f[C]) &= \{b \in {}^*U | * \models (\exists \mathbf{c} \in {}^*\mathbf{C})(\mathbf{b} = {}^*\mathbf{f} \upharpoonright \mathbf{c})\} \\ &= {}^*f[{}^*C]. \quad \blacksquare \end{aligned}$$

An important special case is when $A,B \subseteq S$. In this case, $A \subseteq {}^*A$, $B \subseteq {}^*B$ (by Theorem 7.6). If $a \in A$ and $b = f(a)$, then $\models(\mathbf{b} = \mathbf{f} \upharpoonright \mathbf{a})$ so that by the Transfer Principle $* \models (\mathbf{b} = {}^*\mathbf{f} \upharpoonright \mathbf{a})$, that is, $b = {}^*f(a)$. We have shown that in this case $^*f$ coincides with f on A. In otherwords, $^*f$ simply extends f from A to $^*A$. *In such a case* (i.e., when $\mathrm{dom}(f) \subseteq \mathrm{dom}({}^*f)$ and $^*f$ is simply an extension of f), *we often omit the *, writing f for $^*f$.*

More generally, if r is a relation, $r \subseteq A \times B$, then by Theorem 7.11 and example 3 above, $^*r \subseteq {}^*(A \times B) = {}^*A \times {}^*B$. Again, if $A,B \subseteq S$, $r \subseteq {}^*r$: if $\langle a,b \rangle \in r$, then $\models \langle \mathbf{a},\mathbf{b} \rangle \in \mathbf{r}$, so that $* \models \langle \mathbf{a},\mathbf{b} \rangle \in {}^*\mathbf{r}$ and hence $\langle a,b \rangle \in {}^*r$. As with functions we often omit the * in such a situation, writing r instead of $^*r$.

8. CONCURRENCE, INFINITE INTEGERS, INTERNAL SETS

Although we have developed much of the technique of nonstandard analysis, it is all a bit vacuous so far. We have not even shown that $S \subset W$, merely that $S \subseteq W$. Of course, if $S = W$, then $\hat{S} = \tilde{W} = \hat{W}$, and nothing very exciting is to be expected from the nonstandard universe. And, in fact, it is easy to see that without additional assumptions on the index set I, we are not able to exclude this degenerate case. For example, if $I = \{a\}$ consists of a single element, then the unique ultrafilter on I is $F = \{I\}$. Then $\mu_F(\varnothing) = 0$, $\mu_F(I) = 1$, and, as is readily seen, $W = S$. The same is true as long as I is finite.

In this section, we specify I as a very large set indeed. The key notion is that of concurrence:

DEFINITION. A relation r is called *concurrent in U* (or just concurrent) if $r \in U$ and if whenever $a_1, \ldots, a_n \in \text{dom}(r)$, there is an element b such that $\langle a_i, b \rangle \in r$ for $i = 1, 2, \ldots, n$.

We shall encounter numerous examples of concurrent relations. As a first example, consider for each $k \in N^+$, the relation

$$r_k = \{\langle A,B \rangle | A \subseteq B \subseteq S_k\}$$

Since

$$\begin{aligned} \langle A,B \rangle \in r_k \quad &\text{implies} \quad A,B \in S_{k+1} \\ &\text{implies} \quad \langle A,B \rangle \in S_{k+3}, \end{aligned}$$

we have $r_k \subseteq S_{k+3}$ and $r_k \in S_{k+4}$. Hence certainly $r_k \in U$.

To see that r_k is concurrent, let $A_1, \ldots, A_n \in \text{dom}(r_k)$. That is, let

$$\langle A_1, B_1 \rangle \in r_k, \ldots, \langle A_n, B_n \rangle \in r_h.$$

Let $B = B_1 \cup \cdots \cup B_n$. Since $A_i \subseteq B_i \subseteq S_k$ for $i = 1, \ldots, n$, we have $A_i \subseteq B \subseteq S_k$. That is, $\langle A_i, B \rangle \in r_k$ for $i = 1, \ldots, n$.

The main result about concurrence is:

THEOREM 8.1 (CONCURRENCE THEOREM).[1] Let r be a concurrent relation in U. Then there is an element $b \in {}^*U$ such that $\langle {}^*a, b \rangle \in {}^*r$ for all $a \in \text{dom}(r)$.

For each finite subset A of dom(r), the definition of concurrence yields an element $q_A \in U$ such that

$$\langle a, q_A \rangle \in r \quad \text{for all} \quad a \in A.$$

Intuitively, the Concurrence Theorem yields an element b that is a sort of "limit" of these q_A as A "approaches" dom(r).

We cannot prove the Concurrence Theorem without giving additional information about the index set I and the ultrafilter F used in constructing $\widetilde{W}$. In fact, we now specify I as the set of all functions α such that

(1) α is defined on the set of concurrent relations $r \in U$, and
(2) for each such r, $\alpha(r)$ is a finite subset of dom(r).

We introduce the following notation (which is used only in proving the Concurrence Theorem).

[1] The theorem and proof are from Robinson [15].

For $\alpha,\beta \in I$, we write $\alpha < \beta$ to mean that $\alpha(r) \subseteq \beta(r)$ for each concurrent $r \in U$. Also for $\alpha,\beta \in I$, we write $\gamma = \alpha \vee \beta$ to mean that $\gamma \in I$ is defined by

$$\gamma(r) = \alpha(r) \cup \beta(r).$$

[$\gamma \in I$ since the union of two finite subsets of dom(r) is a finite subset of dom(r).]

Next for each $\alpha \in I$, we write

$$\Gamma_\alpha = \{\beta \in I | \alpha < \beta\}.$$

We then have

LEMMA 1. $\Gamma_\alpha \cap \Gamma_\beta = \Gamma_{\alpha \vee \beta}$.

PROOF.

$$\begin{aligned}
\gamma \in \Gamma_\alpha \cap \Gamma_\beta \quad & \text{if and only if} \quad \alpha < \gamma, \beta < \gamma; \\
& \text{if and only if, for all } r, \quad \alpha(r) \subseteq \gamma(r), \beta(r) \subseteq \gamma(r); \\
& \text{if and only if, for all } r, \quad \alpha(r) \cup \beta(r) \subseteq \gamma(r); \\
& \text{if and only if} \quad \alpha \vee \beta < \gamma; \\
& \text{if and only if} \quad \gamma \in \Gamma_{\alpha \vee \beta}. \quad \blacksquare
\end{aligned}$$

LEMMA 2. Let $G = \{\Gamma_\alpha | \alpha \in I\}$. Then G is a filter basis on I.

PROOF. We verify that G satisfies the three defining conditions for being a filter basis:

(1) Since $\alpha \in \Gamma_\alpha$, we have that for all $\alpha \in I$, $\Gamma_\alpha \neq \varnothing$. Hence $\varnothing \notin G$.
(2) If $\Gamma_\alpha, \Gamma_\beta \in G$, then by Lemma 1, $\Gamma_\alpha \cap \Gamma_\beta = \Gamma_{\alpha \vee \beta} \in G$.
(3) Finally, if we set $\alpha_0(r) = \varnothing$ for all concurrent $r \in U$, we see that $\alpha_0 \in I$ so that $\Gamma_{\alpha_0} \in G$. Hence $G \neq \varnothing$. $\blacksquare$

Using Corollary 2.3 we now stipulate that *F is an ultrafilter on I such that $F \supseteq G$*, thus finally specifying F as well as I.

To summarize:

LEMMA 3. F is an ultrafilter on I such that $\Gamma_\alpha \in F$ for each $\alpha \in I$.

Now, to prove the Concurrence Theorem, let r be a concurrent relation such that $r \in S_k$. Let f map I into U so that for each $\alpha \in I$, $\langle a, f_\alpha \rangle \in r$ for all $a \in \alpha(r)$. Since r is concurrent and $\alpha(r)$ is a finite subset of dom(r), such a map exists. For each α, we then have (because S_k is transitive) $f_\alpha \in S_k$, so that $f \in Z$. We let $b = \bar{f}$, so that $b \in {}^*U$, and plan to show that b satisfies the conclusion of the Concurrence Theorem.

LEMMA 4. For each $a \in \text{dom}(r)$, we have $\langle a, f_\alpha \rangle \in r$ a.e.

PROOF. Let a be some fixed element of dom(r), and let

$$T_a = \{\alpha \in I | \langle a, f_\alpha \rangle \in r\}.$$

We wish to show that $\mu_F(T_a) = 1$, that is, that $T_a \in F$. We proceed to prove this. We define $\beta \in I$ by setting

$$\beta(X) = \begin{cases} \{a\} & \text{if} \quad X = r, \\ \varnothing & \text{if} \quad X \neq r. \end{cases}$$

Now $\Gamma_\beta \in F$ by Lemma 3. Hence it is enough to show that $\Gamma_\beta \subseteq T_a$. But this last is easy to prove. Namely,

$$\begin{aligned} \alpha \in \Gamma_\beta \ \text{implies} \ & \beta < \alpha, \\ \text{implies} \ & \beta(r) \subseteq \alpha(r), \\ \text{implies} \ & a \in \alpha(r), \\ \text{implies} \ & \langle a, f_\alpha \rangle \in r, \\ \text{implies} \ & \alpha \in T_a. \end{aligned}$$ ■

Now we can easily prove the theorem. Let $b = \bar{f}$ and let $h_\alpha = \langle a, f_\alpha \rangle$ for all $\alpha \in I$. By Theorem 4.5, $\bar{h} = \langle \bar{a}, \bar{f} \rangle = \langle {}^*a, b \rangle$. By Lemma 4, $h_\alpha \in r$ a.e., so that by Theorem 4.3 $\bar{h} \in \bar{r} = {}^*r$. Thus $\langle {}^*a, b \rangle \in {}^*r$, and the proof of the Concurrence Theorem is complete. ■

Now let us assume that $N \subseteq S$, that is, that the natural numbers are included among the individuals of the standard universe. Then, of course, $N \in S_1$ so that $N \in \hat{S}$, $\mathscr{P}(N) \in \hat{S}$, and so forth. By Theorem 7.6, $N \subseteq {}^*N$. Consider the relation:

$$L = \{\langle x, y \rangle | x \in N, y \in N, x < y\}.$$

L is clearly concurrent since dom(L) $= N$, and if $a_1, \ldots, a_n \in N$ and b is the largest of $a_1, \ldots, a_n$, then $a_1 L b, \ldots, a_n L b$. By the Concurrence Theorem, there is an element $b \in {}^*U$ such that

$$\langle a, b \rangle \in {}^*L$$

for all $a \in N$ (here ${}^*a = a$ because $a \in N \subseteq S$). Since $L \subseteq N \times N$, we have (recall the discussion in Section 7), ${}^*L \subseteq {}^*N \times {}^*N$. Hence $b \in {}^*N$. Is $b \in N$? If so, we could write ${}^*b = b$ and conclude from $* \models \langle {}^*\mathbf{a}, {}^*\mathbf{b} \rangle \in {}^*\mathbf{L}$ (for all $a \in N$), that $\models \langle \mathbf{a}, \mathbf{b} \rangle \in \mathbf{L}$ (by the Transfer Principle), that is, that $a < b$ for all $a < N$. That is, if $b \in N$, then b is a *largest natural number*. Since such does not exist, we are forced to conclude that $b \in {}^*N - N$. Hence ${}^*N - N \neq \varnothing$. Since $b \in {}^*N - N$, by Theorem 7.6, $b \notin S$; that is, b is a nonstandard individual. In particular, *nonstandard individuals exist*!

This argument really shows that nonstandard individuals exist just as long as S is infinite. For if S is infinite, N can be one-one embedded in S.

Conversely, if S is finite, it is not difficult to show that there are no non-standard elements.

It is natural to write $x < y$ for $\langle x,y \rangle \in {}^*L$ for all $x,y \in {}^*N$ since ${}^*L$ is an extension of the $<$ relation from N to ${}^*N$. We also write $x \leq y$ to mean that either $x < y$ or $x = y$.

The fact that $<$ is a linear ordering of N can be expressed by

$$\models (\forall \mathbf{x} \in \mathbf{N}) \neg (\mathbf{x} < \mathbf{x}),$$
$$\models (\forall \mathbf{x} \in \mathbf{N})(\forall \mathbf{y} \in \mathbf{N})(\forall \mathbf{z} \in \mathbf{N})((\mathbf{x} < \mathbf{y} \,\&\, \mathbf{y} < \mathbf{z}) \rightarrow \mathbf{x} < \mathbf{z}),$$
$$\models (\forall \mathbf{x} \in \mathbf{N})(\forall \mathbf{y} \in \mathbf{N})(\mathbf{x} < \mathbf{y} \vee \mathbf{y} < \mathbf{x} \vee \mathbf{x} = \mathbf{y}).$$

By the Transfer Principle, we conclude that $<$ linearly orders ${}^*N$.

We next show that the elements of ${}^*N - N$ are larger than any natural number.

THEOREM 8.2. If $u \in {}^*N - N$ and $n \in N$, then $n < u$.

PROOF. Suppose otherwise. That is,

$$u \leq n$$

for some $n \in N$. Let n be the *least* such integer. We have

$$\models (\forall \mathbf{x} \in \mathbf{N})(\mathbf{x} \leq \mathbf{0} \rightarrow \mathbf{x} = \mathbf{0}).$$

By the Transfer Principle,

$$* \models (\forall \mathbf{x} \in {}^*\mathbf{N})(\mathbf{x} \leq \mathbf{0} \rightarrow \mathbf{x} = \mathbf{0}).$$

By the semantics of ${}^*\mathscr{L}$, we see that $u \not\leq 0$. Hence $n \neq 0$. So, letting $n = m + 1$, we have $m \in N$, and

$$m < u \leq m + 1.$$

But $u \neq m + 1$ because $m + 1$ is standard. Hence

$$m < u < m + 1.$$

But

$$\models (\forall \mathbf{x} \in \mathbf{N}) \neg (\mathbf{m} < \mathbf{x} < \mathbf{m} + \mathbf{1}).$$

Since $u \in {}^*N$, the Transfer Principle yields

$$* \models \neg (\mathbf{m} < \mathbf{u} < \mathbf{m} + \mathbf{1}).$$

This is a contradiction. ■

This theorem leads us to define:

DEFINITION. For $u \in {}^*N$, we call u *finite* if $u \in N$ and *infinite* (or even an *infinite integer*) if $u \in {}^*N - N$.

Thus the finite elements of $^*N$ are the standard ones, and the infinite elements of $^*N$ are the nonstandard ones. So far, we have only identified one infinite integer. But there are many others. Thus there is a function $P \in U$ such that

$$\text{dom}(P) = N \times N$$

and

$$P(\langle m,n \rangle) = m + n.$$

It is natural to write

$$m + n = {}^*P(\langle m,n \rangle) \in {}^*N$$

for arbitrary $m,n \in {}^*N$. So for $u \in {}^*N - N$, $u + 1$, $u + 2$, $u + 3, \ldots \in {}^*N$. Applying the Transfer Principle to $\models(\forall \mathbf{x} \in \mathbf{N})(\mathbf{x} < \mathbf{x} + \mathbf{1})$, where $\mathbf{x} + \mathbf{1}$ is short for the expression $\mathbf{P} \upharpoonright \langle \mathbf{x},\mathbf{1} \rangle$, we obtain,

$$u < u + 1 < u + 2 < u + 3 < \cdots.$$

Hence $u + 1$, $u + 2$, and so forth are all infinite.

Similarly, given $u,v \in {}^*N$, $u > v$, there is (by the Transfer Principle) an element $z \in {}^*N$ such that $u = v + z$. We write $z = u - v$, so that (Transfer Principle again) $z < u$. Thus we have for $u \in {}^*N - N$,

$$\cdots < u - 3 < u - 2 < u - 1 < u.$$

Each of these elements of $^*N$ must be infinite, since if $u - k$, $k \in N$, then $u = (u - k) + k \in N$.

So beginning with an infinite integer u we obtain a "block" of infinite integers:

$$\cdots < u - 3 < u - 2 < u - 1 < u < u + 1 < u + 2 < u + 3 < \cdots.$$

By using the Transfer Principle, we see that there can be no $v \in {}^*N$ for which

$$u < v < u + 1.$$

However, given a "block," there is another block consisting of even larger infinite integers. For example, there is the integer $u + u$, where $u + k < u + u$ for each $k \in N$. And $v = u + u$ is itself part of the block

$$\cdots < v - 3 < v - 2 < v - 1 < v < v + 1 < v + 2 < \cdots.$$

Of course, $v < v + u < v + v$, and so forth. There are even infinite integers $u \cdot u$ and u^u, and so forth.

Proceeding in the opposite direction, if $u \in {}^*N - N$, either u or $u + 1$ is of the form $v + v$ (by the Transfer Principle). Here v must be infinite. So there is no first "block," since $v < u$. In fact, the ordering of the blocks is dense. For let the block containing v precede the one containing u, that is,

$$v - 2 < v - 1 < v < v + 1 < \cdots < \cdots < u - 2 < u - 1 < u < u + 1 < \cdots$$

Either $u + v$ or $u + v + 1$ can be written $z + z$ where $v + k < z < u - l$ for all $k,l \in N$.

To conclude our discussion: $^*N$ consists of N as an initial segment followed by an ordered set of "blocks." These blocks are densely ordered with no first or last element. Each block is itself order-isomorphic to the integers

$$-3,-2,-1,0,1,2,3,\ldots.$$

Although $^*N - N$ is a nonempty subset of $^*N$, as we have just seen it has no least element (likewise for any "block"). However, any nonempty subset of N does have a least element. Does this situation not contradict the Transfer Principle? It is typical of modern logic to draw a valuable conclusion from such an apparent paradox.

THEOREM 8.3. Let $A \subseteq {}^*N$ be such that $A \in {}^*U$, $A \neq \varnothing$. Then A has a least element.

PROOF. Let $P = \mathscr{P}(N)$. By Theorem 7.12, $A \in {}^*P$. Now, since every nonempty subset of N does have a least element,

$$\models (\forall \mathbf{X} \in \mathbf{P})(\mathbf{X} = \varnothing \vee (\exists \mathbf{m} \in \mathbf{X})(\forall \mathbf{x} \in \mathbf{X})(\mathbf{m} \leq \mathbf{x})).$$

The Transfer Principle yields

$$* \models (\forall \mathbf{X} \in {}^*\mathbf{P})(\mathbf{X} = \varnothing \vee (\exists \mathbf{m} \in \mathbf{X})(\forall \mathbf{x} \in \mathbf{X})(\mathbf{m} \leq \mathbf{x}).$$

Since $A \in {}^*P$ and $A \neq \varnothing$, we conclude from the semantics of $^*\mathscr{L}$ that there is an element $m \in A$ such that $m \leq x$ for $x \in A$. That is, m is the least element of A. ■

COROLLARY 8.4. $^*N - N \notin {}^*U$.

PROOF. Otherwise, it would have a least element.

Sets of $\hat{W}$ that belong to $^*U$ are called *internal*; sets of $\hat{W}$ that are not internal are called *external*: they are literally external to the entire nonstandard universe $^*U$. We have shown that $^*N - N$ is external and hence that *there exist external sets.*

It is very important to be able to demonstrate that specific sets are internal.

THEOREM 8.5 (INTERNALITY THEOREM). Let A be an internal set, and let B be a definable subset of $^*U$. Then $A \cap B$ is internal.

PROOF. We have

$$B = \{b \in {}^*U | * \models \alpha(\mathbf{b})\}$$

for a suitable formula $\alpha = \alpha(\mathbf{x})$ of $^*\mathscr{L}$. Let $\bar{\mathbf{g}}^1, \bar{\mathbf{g}}^2, \ldots, \bar{\mathbf{g}}^n$ be all of the constants that occur in α. Let $\mathbf{y}_1, \ldots, \mathbf{y}_n$ be n variables that don't occur in α, and let

$\gamma = \gamma(\mathbf{x},\mathbf{y}_1,\ldots,\mathbf{y}_n)$ be obtained from α by replacing each $\bar{\mathbf{g}}^i$ by the corresponding variable $\mathbf{y}_i$ at all occurrences. Thus γ is a formula of $\mathscr{L}$ such that $^*\gamma = \gamma$ (because γ contain no constants) and

$$\alpha(\mathbf{x}) = \gamma(\mathbf{x},\bar{\mathbf{g}}^1,\ldots,\bar{\mathbf{g}}^n).$$

Using Łos' Theorem,

$$B = \{\bar{h} \in {}^*U \mid * \models \gamma(\bar{\mathbf{h}},\bar{\mathbf{g}}^1,\ldots,\bar{\mathbf{g}}^n)\}$$
$$= \{\bar{h} \in {}^*U \mid \models \gamma(\mathbf{h}_\delta,\mathbf{g}^1_\delta,\ldots,\mathbf{g}^n_\delta) \text{ a.e.}\}$$

Since A is internal, $A = \bar{g}$ for some $g \in Z$, say $g \in Z_n$. We define k by

$$k_\delta = \{b \in g_\delta \mid \models \gamma(\mathbf{b},\mathbf{g}^1_\delta,\ldots,\mathbf{g}^n_\delta)\}$$

for all $\delta \in I$. Since $g_\delta \in S_n$ a.e. and for each δ, $k_\delta \subseteq g_\delta$, we have $k_\delta \in S_{n+1}$ a.e. and hence $k \in Z$. We shall prove that $\bar{k} = A \cap B$, which will complete the proof.

Now,

$$\bar{h} \in A \quad \text{if and only if} \quad \bar{h} \in \bar{g}$$
$$\text{if and only if} \quad h_\delta \in g_\delta \text{ a.e.}$$

Hence

$$A \cap B = \{\bar{h} \in {}^*U \mid h_\delta \in g_\delta \text{ a.e. and} \models \gamma(\mathbf{h}_\delta,\mathbf{g}^1_\delta,\ldots,\mathbf{g}^n_\delta) \text{ a.e.}\}$$
$$= \{\bar{h} \in {}^*U \mid h_\delta \in k_\delta \text{ a.e.}\}$$
$$= \{\bar{h} \in {}^*U \mid \bar{h} \in \bar{k}\} = \bar{k}. \quad \blacksquare$$

COROLLARY 8.6. If $B \subseteq A$ where A is internal and B is definable in $^*U$, then B is internal.

PROOF. $B = A \cap B$. $\blacksquare$

As an application, we now have

COROLLARY 8.7. N is external.

PROOF. Suppose that N were internal, that is, $N \in {}^*U$. Then $^*N - N$ would be definable by

$$^*N - N = \{b \in {}^*U \mid * \models (\mathbf{b} \in {}^*\mathbf{N}\ \&\ \neg(\mathbf{b} \in \mathbf{N}))\}.$$

(The internality of N is being used in assuming the existence of a constant $\mathbf{N}$ in $^*\mathscr{L}$.) Since $^*N$ is internal and $^*N - N \subseteq {}^*N$, it would follow that $^*N - N$ is internal, contradicting Corollary 8.4. $\blacksquare$

THEOREM 8.8. If A and B are internal, so is $A \times B$.

PROOF. Using Theorem 7.10, let $A,B \in {}^*S_i$. By Theorem 3.1,

$$\models (\forall \mathbf{X} \in \mathbf{S}_i)(\forall \mathbf{Y} \in \mathbf{S}_i)(\exists \mathbf{Z} \in \mathbf{S}_{i+3})(\mathbf{Z} = \mathbf{X} \times \mathbf{Y}).$$

Using the Transfer Principle,

$$* \models (\forall \mathbf{X} \in {}^*\mathbf{S}_i)(\forall \mathbf{Y} \in {}^*\mathbf{S}_i)(\exists \mathbf{Z} \in {}^*\mathbf{S}_{i+3})(\mathbf{Z} = \mathbf{X} \times \mathbf{Y}).$$

By the semantics of ${}^*\mathscr{L}$, we have $C \in {}^*S_{i+3}$ such that $C = A \times B$. But using Theorem 7.10 again, we see that C is internal. ■

THEOREM 8.9. (INTERNAL FUNCTION THEOREM). Let f map A into B where A,B are internal subsets of ${}^*U$. Let $\mu = \mu(\mathbf{x})$ be a term of ${}^*\mathscr{L}$ such that

$$f(a) = |\mu(\mathbf{a})|_*$$

for each $a \in A$. Then f is internal.

PROOF. We have

$$f = \{c \in {}^*U | * \models (\exists \mathbf{x} \in \mathbf{A})(\exists \mathbf{y} \in \mathbf{B})(\mathbf{c} = \langle \mathbf{x},\mathbf{y} \rangle \;\&\; \mathbf{y} = \mu(\mathbf{x}))\}$$

Therefore, f is a definable subset of ${}^*U$. Also $f \subseteq A \times B$, and by Theorem 8.8, $A \times B$ is internal. By Corollary 8.6, f is an internal set. ■

9. RECAPITULATION

We have now completed our development of the basic techniques of nonstandard analysis, and it is time to pause to take stock. Much of the machinery used in this chapter was only needed for our basic constructions and is not used in the remainder of the book.

Let us recapitulate the key ideas and results that a reader should bear in mind:

(1) We have introduced two universes: the standard universe $U = \hat{S}$, which is simply the superstructure built on a suitable set of individuals S (an infinite set if we are to avoid triviality), and the nonstandard universe ${}^*U = \tilde{W}$, which consists of a set of individuals $W \supseteq S$ ($W \supset S$ if S is infinite) and certain of the sets from the superstructure $\hat{W}$.
(2) The sets of $\hat{W}$ consist of the *internal sets*, which are in ${}^*U$, and the external sets, which are not.
(3) A map * exists mapping U into ${}^*U$. Elements of ${}^*U$ of the form ${}^*x, x \in U$ (i.e., elements of the range of the * map) are called *standard*; the remaining elements are called nonstandard.
(4) Languages $\mathscr{L}$, ${}^*\mathscr{L}$ are available for making assertions about, and defining subsets of U and ${}^*U$, respectively. The * map induces a map of sentences of $\mathscr{L}$ into those of ${}^*\mathscr{L}$, and the all-important *Transfer Principle* holds: α is true in U if and only if ${}^*\alpha$ is true in ${}^*U$.
(5) Relations r in S that are concurrent (i.e., such that $A \subseteq \text{dom}(r)$ and A finite implies the existence of b such that $\langle a,b \rangle \in r$ for all $a \in A$) lead to

the existence of "limiting" elements in $*U$, that is, elements $b \in *U$ such that $\langle *a,b \rangle \in *r$ for all $a \in \text{dom}(r)$. This provides a kind of universal all-at-once compactification and completion in a sense that will be clear later.

(6) There are external sets; even the ordinary set of natural numbers is an external set. Nevertheless, any set that is bounded by an internal set and can be defined in $*\mathscr{L}$ is internal.

The reader who remembers these key points will do very well in what follows. In particular, it is now quite all right to entirely forget how the nonstandard universe was defined and to banish ultrafilters from our consciousness.

EXERCISES

1. Let F be an ultrafilter on a given nonempty set I. Let $A,B \subseteq I$ where $A \cap B = \varnothing$. Show that

$$\mu_F(A \cup B) = \mu_F(A) + \mu_F(B).$$

   (Of course it is because of this property that μ_F is called a measure.)
2. Show directly that N is external by using a sentence of $\mathscr{L}$ that expresses the principle of mathematical induction.
3. Prove Theorem 8.8 using the Internality Theorem.
4. Let σ be a function such that $\text{dom}(\sigma) = N$ and for each $n \in N$, $\sigma(n) = \{m \in N | m \leq n\}$. Show that for $n \in *N$, $*\sigma(n) = \{m \in *N | m \leq n\}$.
5. Show that each "block" in $*N$ is an external set.
6. Let $r = \{\langle a,b \rangle | \models \alpha(\mathbf{a},\mathbf{b})\}$ where $r \in U$. Show that $*r = \{\langle a,b \rangle | \models *\alpha(\mathbf{a},\mathbf{b})\}$
7. Show that $*\mathscr{P}(A)$, where $A \in U$, is the set of all internal subsets of $*A$.

2
REAL NUMBERS AND HYPERREAL NUMBERS

1. ORDERED FIELDS

So far, we have applied nonstandard analysis only to the natural numbers N. As our next example, we apply nonstandard methods to ordered fields. The same construction then serves for two distinct purposes: (1) to give a neat elegant nonstandard substitute for the classical "construction" of the real numbers, and (2) to work out the structure of $^*R$ where R is the set of real numbers.

In this section, we use only standard methods to obtain the information about ordered fields that we need. Nonstandard methods are then applied in Section 2.

Let $D \subseteq S$, let 0, 1 be elements of D, and let $+, \cdot$ each map $D \times D$ into D. As usual, we write $a + b$ for $+ \upharpoonright \langle a,b \rangle$ and $a \cdot b$ or even ab for $\cdot \upharpoonright \langle a,b \rangle$. Consider the sentences Φ listed below:

$\Phi 1.$ $(\forall \mathbf{x} \in \mathbf{D})(\forall \mathbf{y} \in \mathbf{D})((\mathbf{x} + \mathbf{y}) = (\mathbf{y} + \mathbf{x}))$.
$\Phi 2.$ $(\forall \mathbf{x} \in \mathbf{D})(\forall \mathbf{y} \in \mathbf{D})(\forall \mathbf{z} \in \mathbf{D})((\mathbf{x} + (\mathbf{y} + \mathbf{z})) = ((\mathbf{x} + \mathbf{y}) + \mathbf{z}))$.
$\Phi 3.$ $(\forall \mathbf{x} \in \mathbf{D})((\mathbf{x} + \mathbf{0}) = \mathbf{x})$.
$\Phi 4.$ $(\forall \mathbf{x} \in \mathbf{D})(\exists \mathbf{y} \in \mathbf{D})((\mathbf{x} + \mathbf{y}) = \mathbf{0})$.
$\Phi 5.$ $(\forall \mathbf{x} \in \mathbf{D})(\forall \mathbf{y} \in \mathbf{D})((\mathbf{x} \cdot \mathbf{y}) = (\mathbf{y} \cdot \mathbf{x}))$.
$\Phi 6.$ $(\forall \mathbf{x} \in \mathbf{D})(\forall \mathbf{y} \in \mathbf{D})(\forall \mathbf{z} \in \mathbf{D})((\mathbf{x} \cdot (\mathbf{y} \cdot \mathbf{z})) = ((\mathbf{x} \cdot \mathbf{y}) \cdot \mathbf{z}))$.
$\Phi 7.$ $(\forall \mathbf{x} \in \mathbf{D})((\mathbf{x} \cdot \mathbf{1}) = \mathbf{x})$.
$\Phi 8.$ $(\forall \mathbf{x} \in \mathbf{D})(\neg(\mathbf{x} = \mathbf{0}) \rightarrow (\exists \mathbf{y} \in \mathbf{D})((\mathbf{x} \cdot \mathbf{y}) = \mathbf{1}))$.
$\Phi 9.$ $\neg(\mathbf{0} = \mathbf{1})$.
$\Phi 10.$ $(\forall \mathbf{x} \in \mathbf{D})(\forall \mathbf{y} \in \mathbf{D})(\forall \mathbf{z} \in \mathbf{D})((\mathbf{x} \cdot (\mathbf{y} + \mathbf{z})) = ((\mathbf{x} \cdot \mathbf{y}) + (\mathbf{x} \cdot \mathbf{z})))$.

To say that the sentences Φ are true in $\mathscr{L}$ is just to say that D forms a *field* in the usual sense with operations $+, \cdot$ and corresponding identity elements 0, 1. In addition, let $<$ be a relation on D, and consider the sentences:

$\Omega 1.$ $(\forall \mathbf{x} \in \mathbf{D})\neg(\mathbf{x} < \mathbf{x})$.
$\Omega 2.$ $(\forall \mathbf{x} \in \mathbf{D})(\forall \mathbf{y} \in \mathbf{D})(\forall \mathbf{z} \in \mathbf{D})((\mathbf{x} < \mathbf{y} \;\&\; \mathbf{y} < \mathbf{z}) \rightarrow \mathbf{x} < \mathbf{z})$.

$\Omega 3.$ $(\forall \mathbf{x} \in \mathbf{D})(\forall \mathbf{y} \in \mathbf{D})(\mathbf{x} < \mathbf{y} \vee \mathbf{y} < \mathbf{x} \vee \mathbf{x} = \mathbf{y})$.
$\Omega 4.$ $(\forall \mathbf{x} \in \mathbf{D})(\forall \mathbf{y} \in \mathbf{D})(\forall \mathbf{z} \in \mathbf{D})(\mathbf{x} < \mathbf{y} \rightarrow \mathbf{x} + \mathbf{z} < \mathbf{y} + \mathbf{z})$.
$\Omega 5.$ $(\forall \mathbf{x} \in \mathbf{D})(\forall \mathbf{y} \in \mathbf{D})(\forall \mathbf{z} \in \mathbf{D})((\mathbf{x} < \mathbf{y} \;\&\; \mathbf{0} < \mathbf{z}) \rightarrow \mathbf{x} \cdot \mathbf{z} < \mathbf{y} \cdot \mathbf{z})$.

Then the sentences Φ, Ω are all true in $\mathscr{L}$ just in case D forms an *ordered field* under $<$. (The reader is free to regard the previous sentence as a definition.) Below, we assume that D is an ordered field. As usual, we write $x > y$ for $y < x$, $x \leq y$ for $x < y$ or $x = y$. If $x < y$ and $u < v$, then

$$x + u < y + u < y + v.$$

LEMMA 1. $0 < 1$. For all $x \in D$, $x < x + 1$.

PROOF. By $\Phi 9$, $0 \neq 1$. If it were not true that $0 < 1$, then by $\Omega 3$ we would have that $1 < 0$. By $\Omega 4$, $0 < -1$. Using $\Omega 5$, we can multiply both sides of this inequality by -1 to obtain $0 < 1$.

Finally, we have $x = x + 0 < x + 1$, which completes the proof. ∎

Thus

$$0 < 1 < 1 + 1 < 1 + 1 + 1 < \cdots.$$

Hence the map

$$\begin{array}{cccccc} 0 & 1 & 2 & 3 & 4 & \cdots \\ \downarrow & \downarrow & \downarrow & \downarrow & \downarrow & \\ 0 & 1 & 1+1 & 1+1+1 & 1+1+1+1 & \cdots \end{array}$$

is an order isomorphism of N into D. By $\Phi 1$–3, $\Phi 5$–7, this map also preserves $+$ and $\cdot$. Therefore, it is natural to regard this map as an identification and to think of N as embedded in D; henceforth we therefore can assume that $N \subset D$.

We write Q for the set of rational numbers. Of course, Q is an ordered field. Moreover, since each element of Q can be written as a/b or $-(a/b)$ where $a,b \in N$, the embedding of N in D induces an embedding of Q in D, which is in fact an isomorphism. We regard this embedding too as an identification, so that henceforth we have $Q \subseteq D$.

LEMMA 2. $x < y$ implies $-y < -x$.

PROOF. If $-x = -y$, then $y = x < y$, which contradicts $\Omega 1$. If $-x < -y$, then $0 = x + (-x) < y + (-y) = 0$, which also contradicts $\Omega 1$. The result follows from $\Omega 3$. ∎

DEFINITION. For all $x \in D$,

$$|x| = \max(x, -x).$$

The following is easily established in the usual way, for example, by considering various cases.

LEMMA 3. $|x + y| \leq |x| + |y|$ and $|x \cdot y| = |x| \cdot |y|$.

DEFINITION. D is called *Archimedean* if for each $x \in D$ there exists an $n \in N$ such that $x < n$. Otherwise, D is called *non-Archimedean*.

The rational numbers provide an example of an Archimedean ordered field. We now obtain some basic properties of Archimedean fields.

LEMMA 4. Let D be Archimedean. If $x,y \in D$, $x,y > 0$, and $y - x > 1$, then for some $n \in N$, $x < n < y$.

PROOF. Since D is Archimedean, there exists a natural number $n > x$. Let n be the least such. Since $x > 0$, $n \neq 0$. We show that $n < y$.

By the minimality of n, $n - 1 \leq x$. By hypothesis, $1 < y - x$. Hence $n = (n - 1) + 1 < x + (y - x) = y$. ■

THEOREM 1.1 (DENSITY). If D is an Archimedean field and $x,y \in D$ where $x < y$, then there exists an $r \in Q$ such that $x < r < y$.

PROOF.

CASE 1

$x > 0$. Consider the positive element $1/(y - x)$ of D. Since D is Archimedean, there exists some $m \in N$ such that

$$\frac{1}{y - x} < m.$$

It follows that

$$my - mx > 1.$$

By Lemma 4, there exists an $n \in N$ such that

$$mx < n < my.$$

Since $1/m > 0$,

$$x < \frac{n}{m} < y.$$

CASE 2

$x = 0$. Choose $n \in N$, $n > 1/y$. Then $0 = x < 1/n < y$.

CASE 3

$x < 0 < y$. Choose $r = 0$.

CASE 4

$y \le 0$. Then, by Lemma 2, $0 \le -y < -x$. By Case 1 (or 2), there exists $r \in Q$, such that $-y < r < -x$, that is, using Lemma 2 once again, we have $x < -r < y$. ■

The element $l \in D$ is an *upper bound* of the nonempty set $A \subseteq D$ if $x \le l$ for all $x \in A$. If, in addition, no $m \in D$ for which $m < l$ is an upper bound of A, l is called the *least upper bound* of A.

DEFINITION. D is *complete* if every nonempty subset of D that has an upper bound has a least upper bound.

THEOREM 1.2. A complete ordered field is Archimedean.

PROOF. Suppose D is a complete ordered field that is not Archimedean. Then there is an $x \in D$ such that $x > n$ for all $n \in N$. So x is an upper bound of N. Hence since D is complete, N has a least upper bound l.

Since l is a *least* upper bound of N, $l - 1$ is not an upper bound of N, that is, $l - 1 < m$ for some $m \in N$, so that $l < m + 1$. But this is a contradiction. ■

THEOREM 1.3. Let D, $\bar{D}$ be complete ordered fields. Then there is a unique one-one map ψ of D onto $\bar{D}$ such that $\psi(q) = q$ for $q \in Q$ and such that for $x,y \in D$, $x < y$ if and only if $\psi(x) < \psi(y)$.

PROOF. For $x \in D$, let $\psi(x)$ be the least upper bound in $\bar{D}$ of

$$S_x = \{r \in Q | r < x\}.$$

(S_x has an upper bound in $\bar{D}$ since for some $n \in N$, $x < n$ because D is Archimedean.)

For $q \in Q$, $\psi(q) \le q$, since q is an upper bound of S_q. Moreover, if $x \in \bar{D}$ and $x < q$, then by Density there is an $r \in Q$, $x < r < q$, so that $r \in S_q$ and x is not an upper bound of S_q. Hence $\psi(q) = q$.

If $x < y$, then by Density, there are $q,r \in Q$ such that $x < q < r < y$. Then q is an upper bound of S_x and $r \in S_y$. So $\psi(x) \le q < r \le \psi(y)$. Conversely, if $\psi(x) < \psi(y)$ then $x < y$, since $y < x$ would imply $\psi(y) < \psi(x)$.

Let $\bar{\psi}$ be another map with the required properties. If $r \in S_x$, then $r < x$, so $r = \bar{\psi}(r) < \bar{\psi}(x)$. That is, $\bar{\psi}(x)$ is an upper bound of S_x in $\bar{D}$. We shall see that $\bar{\psi}(x)$ is the *least* upper bound of S_x in $\bar{D}$, so that $\bar{\psi}(x) = \psi(x)$. Let $\bar{\psi}(y) < \bar{\psi}(x)$. Then, as above, $y < x$, and we may choose $r \in Q$ with $y < r < x$. Thus $r \in S_x$ and $\bar{\psi}(y) < \bar{\psi}(r) = r$, and hence $\bar{\psi}(y)$ cannot be an upper bound of S_x. ■

Later, we see that $\psi(x)$ is, in fact, an isomorphism of D onto $\bar{D}$.

Next we turn to non-Archimedean fields. For any ordered field D we define:

DEFINITION. $F = \{x \in D \mid |x| < n \text{ for some } n \in N\}$. If $x \in F$, x is called *finite*; if $x \in D - F$, x is called *infinite*.

DEFINITION. $I = \{x \in D \mid x = 0 \text{ or } (1/x) \in D - F\}$. The elements of I are called *infinitesimal*.

The case where D is Archimedean is thus characterized by either of the equivalent conditions:

$$D = F; \qquad I = \{0\}.$$

LEMMA 5. $x \in I$ if and only if $|x| \leq 1/n$ for all $n \in N^+$.

PROOF. $x \in I$ if and only if either $x = 0$ or $1/x$ is infinite, that is, if and only if either $x = 0$ or $|1/x| \geq n$ for all $n \in N$, or finally if and only if $|x| \leq 1/n$ for all $n \in N$. ■

Note, in particular, that for $x \in I$, $|x| \leq 1$; hence $I \subseteq F$.

THEOREM 1.4. If $x \in F$ and $y \in F$, then $x + y \in F$ and $x \cdot y \in F$. (That is, F is a subring of D.)

PROOF. Let $x,y \in F$, $|x| < m$, $|y| < n$, $m,n \in N$. Then $|x + y| \leq |x| + |y| < m + n$, and $|xy| = |x| \cdot |y| < m \cdot n$. Thus $x + y \in F$ and $x \cdot y \in F$. ■

THEOREM 1.5. We have

(1) $x,y \in I$ implies $x + y \in I$;
(2) $x \in I$, $y \in F$ implies $x \cdot y \in I$.

(That is, I is an ideal in F.)

PROOF. (1) For all $n \in N^+$,

$$|x + y| \leq |x| + |y| \leq \frac{1}{2n} + \frac{1}{2n} = \frac{1}{n};$$

Hence $x + y \in I$.

(2) Since $y \in F$, there is some $m \in N$ such that $|y| < m$. Hence for all $n \in N^+$,

$$|xy| = |x| \cdot |y| < m|x| \leq m \cdot \frac{1}{mn} = \frac{1}{n}.$$

Thus $xy \in I$. ■

DEFINITION. We write $x \approx y$ if $x,y \in D$ and $x - y \in I$. In this case, we say that x *is infinitesimally near* y or just x *is near* y. Otherwise, we write $x \not\approx y$.

LEMMA 6. $\approx$ is an equivalence relation on D and hence on F.

PROOF. $x - x = 0 \in I$, so $x \approx x$. Since $(-1) \in F$, $x - y \in I$ implies that $y - x = (-1)(x - y) \in I$. That is, $x \approx y$ implies $y \approx x$. Finally, if $x - y \in I$ and $y - z \in I$, then $x - z = (x - y) + (y - z) \in I$. ■

Since I is an ideal in the ring F, we can construct the quotient ring F/I. Then there is a natural homomorphism of F onto F/I with kernel I. For each $x \in F$, we write $^\circ x$ for the corresponding element of F/I under this homomorphism. Then for $x,y \in F$, that is, $^\circ x = {}^\circ y$ if and only if $x - y \in I$, that is, if and only if $x \approx y$.

Since $^\circ$ is a homomorphism, we have the homomorphism equations:

$$^\circ(x + y) = {}^\circ x + {}^\circ y,$$
$$^\circ(x \cdot y) = {}^\circ x \cdot {}^\circ y.$$

An alternative way of writing this is: For $x,y,x_1,y_1 \in F$, $x \approx x_1$ and $y \approx y_1$ implies $x + y \approx x_1 + y_1$ and $xy \approx x_1 y_1$.

LEMMA 7. If $x \in D - I$, then $1/x \in F$.

PROOF. Since $x \notin I$, for some $n \in N^+$, $|x| > 1/n$. Then $|1/x| < n$. ■

THEOREM 1.6. F/I is a field.

PROOF. Let $^\circ x \in F/I$, $^\circ x \neq 0$. Thus $x \notin I$. By Lemma 7, $1/x \in F$. Then

$$^\circ x \cdot {}^\circ\!\left(\frac{1}{x}\right) = {}^\circ\!\left(x \cdot \frac{1}{x}\right) = {}^\circ 1 = 1.$$

Thus $^\circ x$ has a reciprocal in F/I. ■

We next examine the ordering induced in the field F/I by the ordering of F.

LEMMA 8. If $x \in F - I$, $x > 0$, and $t \in I$, then $x + t > 0$.

PROOF. Since x is positive and noninfinitesimal, for some $n \in N^+$,

$$x > \frac{1}{n}.$$

We need only consider the case $t < 0$. Since t is infinitesimal $|t| \leq 1/n$; hence

$$t \geq -\frac{1}{n}.$$

Adding the displayed inequalities, we have $x + t > 0$. ■

Since r is not an upper bound of C and, being rational, is in D, we have $r \in A$. But this implies $r \leq c$, a contradiction.

c is the least upper bound of C: For, suppose $d < c$ and d is an upper bound of C. Choose $r \in Q$, with $d < r < c$. Since $r \in D$ and r is likewise an upper bound of C, $r \in B$. So $c \leq r$, a contradiction. ■

As we shall see, the Transfer Principle is often used to replace an "arbitrarily fine" approximation in U by an "infinitesimal" approximation in $^*U$. We conclude this section with a first example:

THEOREM 2.6. For each $x \in D$, there is a $q \in {}^*Q$ such that $x \approx q$.

PROOF. By the density of Q in D (Theorem 1.1), we have, for each $x \in D$,

$$\models (\forall \mathbf{n} \in \mathbf{N}^+)(\exists \mathbf{q} \in \mathbf{Q})(|\mathbf{x} - \mathbf{q}| < 1/\mathbf{n}).$$

By the Transfer Principle:

$$* \models (\forall \mathbf{n} \in {}^*\mathbf{N}^+)(\exists \mathbf{q} \in {}^*\mathbf{Q})(|\mathbf{x} - \mathbf{q}| < 1/\mathbf{n}).$$

Let $n \in {}^*N - N$. Then by the semantics of $^*\mathscr{L}$, there is a $q \in {}^*Q$ such that $|x - q| < 1/n$, that is, $x \approx q$. ■

3. THE REAL NUMBERS

In Section 2, $D \subseteq S$ was an unspecified Archimedean ordered field. Now we specify D as the field Q of rational numbers. In this case, we write $R = F/I$. In the general case, D was isomorphically embedded in F/I; thus we have $Q \subseteq R$. Noting Theorem 2.5, we have

THEOREM 3.1. There is a complete ordered field R.

Now what is usually called the "construction" of the real number system amounts precisely to proving the existence and uniqueness (up to isomorphism) of a complete ordered field. Standard proofs of Theorem 3.1 require the "construction" of the real numbers (via Dedekind cuts, Cauchy sequences, or the like), *definition* of $+, \cdot, <$, and a rather tedious verification of the ordered-field properties. The Transfer Principle has done all of this for us at one blow. Only the completeness property remained to be proved, and here the Concurrence Theorem came to our aid.

To complete our discussion, let D, $\bar{D}$ be two complete ordered fields. By Theorem 1.3, there is a (unique) order-preserving map ψ of D onto $\bar{D}$ that is the identity on Q. We show that ψ is in fact an isomorphism of D onto $\bar{D}$.

For this purpose, let us assume that D, $\bar{D}$ are themselves subsets of S, the set of standard individuals. (Naturally, it doesn't matter in the least that we have only located complete ordered fields in the nonstandard universe. Once having proved their existence, there is nothing to prevent us from importing them into the standard universe). Then $^*D$, $^*\bar{D}$ are (non-Archimedean) ordered fields, and $^*\psi$ maps $^*D$ onto $^*\bar{D}$. Since $D \subseteq {}^*D$, $\bar{D} \subseteq {}^*\bar{D}$, we write ψ for $^*\psi$. Applying the Transfer Principle to Theorem 1.3, we have

(1) for $x,y \in {}^*D$, $x < y$, if and only if $\psi(x) < \psi(y)$,
(2) for $q \in {}^*Q$, $\psi(q) = q$.

LEMMA 1. If $x \in D$ and $q \in {}^*Q$, then

$$x \approx q \quad \text{if and only if} \quad \psi(x) \approx q.$$

PROOF.

$$\begin{aligned} x \approx q \quad &\text{if and only if} \quad q - \frac{1}{n} < x < q + \frac{1}{n} \quad \text{for all} \quad n \in N^+, \\ &\text{if and only if} \quad q - \frac{1}{n} < \psi(x) < q + \frac{1}{n} \quad \text{for all} \quad n \in N^+, \\ &\text{if and only if} \quad \psi(x) \approx q. \quad \blacksquare \end{aligned}$$

LEMMA 2. For $x,y \in D$,

$$\psi(x + y) = \psi(x) + \psi(y),$$

and

$$\psi(xy) = \psi(x)\psi(y).$$

PROOF. By Theorem 2.6, we can choose $q,r \in {}^*Q$ such that

$$x \approx q, \qquad y \approx r.$$

Then

$$x + y \approx q + r, \qquad xy \approx qr.$$

By Lemma 1,

$$\begin{aligned} \psi(x) &\approx q, & \psi(y) &\approx r; \\ \psi(x + y) &\approx q + r, & \psi(xy) &\approx qr. \end{aligned}$$

so

$$\begin{aligned} \psi(x + y) &\approx \psi(x) + \psi(y), \\ \psi(xy) &\approx \psi(x) \cdot \psi(y). \end{aligned}$$

Since both sides of these last relations are standard (i.e., in D), we can replace the "$\approx$" by "$=$". $\blacksquare$

Combining Lemma 2 with Theorem 3.1, we have

THEOREM 3.2. There is a complete ordered field R that is unique up to isomorphism.

4. THE HYPERREAL NUMBERS

In Section 2, we let D be an Archimedean ordered field, $D \subseteq S$, and studied $^*D$. As we saw in Section 3, by specifying $D = Q$, we obtain the real number system R.

In this section, we set $D = R$ and study the non-Archimedean ordered field (Theorem 2.1) $^*R$, called the *hyperreal numbers*. Since $Q \subset R$, $^*Q \subseteq {}^*R$. Also, $R \subset {}^*R$ where the inclusion is proper because $^*R$ contains the infinite integers.

Since $^*R$ is non-Archimedean, the considerations of Section 1 apply, and we can speak of F, the finite elements of $^*R$, and I, the infinitesimal elements of $^*R$. Moreover, we have the homomorphic map $^\circ$ of $^*R$ into the ordered Archimedean field F/I (Theorems 1.8, 1.9). Continuing into Section 2, we see that F/I is a complete ordered field (Theorem 2.5). Hence by the considerations of Section 3, F/I is isomorphic (and uniquely so) to R! So we simply identify F/I with R.

To recapitulate, we begin with $R \subseteq S$. In the nonstandard universe we have the non-Archimedean field $^*R$. And there is a homomorphic map $^\circ$ of the ring of finite elements of $^*R$ into R. For x a finite hyperreal number, $^\circ x$ is called the *standard part of* x. The following theorems summarize previous results in the present context:

THEOREM 4.1. For $x,y \in F$, we have

$$^\circ(x + y) = {}^\circ x + {}^\circ y,$$
$$^\circ(xy) = {}^\circ x \cdot {}^\circ y,$$
$$x \leq y \quad \text{if and only if} \quad {}^\circ x \leq {}^\circ y.$$

If $x,y \in R$, then $^\circ x = x$.

THEOREM 4.2. Let $x,y,x_1,y_1 \in F$. Then if $x \approx x_1$, $y \approx y_1$, we have

$$x + y \approx x_1 + y_1,$$
$$xy \approx x_1 \cdot y_1.$$

We also have the following infinitesimal calculus:

THEOREM 4.3. We have

(1) $x \approx 0$, $y \approx 0$ implies $x + y \approx 0$;
(2) $x \approx 0$, $y \in F$ implies $xy \approx 0$;
(3) $x \in F$, $x \neq 0$ implies $1/x \in F$.

PROOF. (1) and (2) are just Theorem 1.5 applied to the present case. For (3), compare with Lemma 7 of Section 1. ■

COROLLARY 4.4. If $x \in F$, $x \neq 0$, $x \approx y$, then $1/x \approx 1/y$.

PROOF.

$$\frac{1}{x} - \frac{1}{y} = \frac{1}{x} \cdot \frac{1}{y} \cdot (x - y).$$

The result follows from Theorem 4.3, (2) and (3). ■

The relation $\approx$ (defined by $x \approx y$ if and only if $x - y \in I$) is an equivalence relation on all of $*R$ (not merely on F) (cf. Lemma 6 of Section 1). For $x \in F$, the equivalence class containing x contains a unique real number, $°x$, the standard part of x.

Theorem 2.6 may now be written:

THEOREM 4.5. For every real number x, there is a $q \in *Q$ such that $x \approx q$.

5. REAL SEQUENCES AND FUNCTIONS

In this section, we apply the hyperreal numbers to show how some typical results from elementary real analysis can be obtained by nonstandard methods. All we really use is that $*R \supset R$ and that the Transfer Principle holds. Our aim is not to be systematic, since we examine the matter later in a more general setting. From now on we write R^+ for the set of *positive real numbers*. Then by the Transfer Principle, $*R^+ = *(R^+)$ is the set of hyperreal numbers x such that $x > 0$.

Let $\{s_n | n \in N^+\}$ be a sequence of real numbers; that is, let s map N^+ into R. Then $*s$ maps $*N^+$ (of course, $*N^+ = *N - \{0\}$) into $*R$. As is natural, we write s_n for $*s(n)$ even when n is infinite. Then we have

THEOREM 5.1. $s_n \to L$ if and only if $s_n \approx L$ for all infinite n.

PROOF. Let $s_n \to L$, and let us choose some $\varepsilon \in R^+$. Corresponding to this ε there exists some $n_0 \in N$ such that

$$\models (\forall \mathbf{n} \in \mathbf{N})(\mathbf{n} > \mathbf{n}_0 \to |\mathbf{s}_n - \mathbf{L}| < \varepsilon).$$

Using the Transfer Principle, we see that for any $n \in *N$ for which $n > n_0$, we have

$$|s_n - L| < \varepsilon.$$

Since n_0 is finite, this inequality holds for all $n \in *N - N$. But ε was chosen to be *any* positive real number. Hence we can conclude that $|s_n - L| \approx 0$, that is, $s_n \approx L$ for any infinite integer n.

Conversely, let $s_n \approx L$ for all $n \in *N - N$, and again choose $\varepsilon \in R^+$. Since $s_n \approx L$ for all infinite n,

$$|s_n - L| < \varepsilon$$

for all infinite n. In particular, if n_0 is some fixed infinite integer, $|s_n - L| < \varepsilon$ for all $n > n_0$. Hence

$$* \models (\exists \mathbf{n}_0 \in {}^*\mathbf{N})(\forall \mathbf{n} \in {}^*\mathbf{N})(\mathbf{n} > \mathbf{n}_0 \rightarrow |\mathbf{s}_n - \mathbf{L}| < \varepsilon).$$

Using the Transfer Principle, we see that the standard definition of $s_n \rightarrow L$ is satisfied. ■

COROLLARY 5.2. Let $s_n \rightarrow L$, $t_n \rightarrow M$. Then

$$s_n + t_n \rightarrow L + M,$$
$$s_n t_n \rightarrow LM,$$

and if $M \neq 0$,

$$\frac{s_n}{t_n} \rightarrow \frac{L}{M}.$$

PROOF. For n infinite, $s_n \approx L$, $t_n \approx M$. So, by Theorem 4.3, $s_n + t_n \approx L + M$, $s_n t_n \approx LM$, which gives the first two results.

Since M is real, it is finite. Hence t_n is finite even for infinite n. By Corollary 4.4, $1/t_n \approx 1/M$. So, $s_n/t_n = s_n(1/t_n) \approx L(1/M) = L/M$. ■

Recall that x is said to be a *limit point* of $\{s_n | n \in N^+\}$ if for every $\varepsilon \in R^+$, the inequality $|s_n - x| < \varepsilon$ is satisfied for infinitely many n. We have

THEOREM 5.3. x is a limit point of $\{s_n | n \in N^+\}$ if and only if $s_n \approx x$ for *some* infinite integer n.

PROOF. If L is a limit point of $\{s_n\}$, then

$$\models (\forall \varepsilon \in \mathbf{R}^+)(\forall \mathbf{n} \in \mathbf{N})(\exists \mathbf{m} \in \mathbf{N})(\mathbf{m} > \mathbf{n} \ \& \ |\mathbf{s}_\mathbf{m} - \mathbf{x}| < \varepsilon).$$

By the Transfer Principle,

$$* \models (\forall \varepsilon \in {}^*\mathbf{R}^+)(\forall \mathbf{n} \in {}^*\mathbf{N})(\exists \mathbf{m} \in {}^*\mathbf{N})(\mathbf{m} > \mathbf{n} \ \& \ |\mathbf{s}_\mathbf{m} - \mathbf{x}| < \varepsilon).$$

Choose ε infinitesimal and n infinite. We conclude that there is an $m > n$ (hence m is also infinite) such that $|s_m - x| < \varepsilon$. Since $\varepsilon \approx 0$, $s_m \approx x$.

Conversely, let $s_n \approx x$ for some fixed infinite n. Let $\varepsilon \in R^+$ so that we have $|s_n - x| < \varepsilon$. Let $m \in N$ be fixed. Since $n > m$,

$$* \models (\exists \mathbf{n} \in {}^*\mathbf{N})(\mathbf{n} > \mathbf{m} \ \& \ |\mathbf{s}_\mathbf{n} - \mathbf{x}| < \varepsilon).$$

Using the Transfer Principle, we obtain an $n \in N$ such that $n > m$ and $|s_n - x| < \varepsilon$. Since $m \in N$ was arbitrary, this last inequality holds for infinitely many n. ■

As usual, we write $[a,b] = \{x \in R | a \leq x \leq b\}$ where $a,b \in R$ and $a \leq b$.

THEOREM 5.4. Let f map $[a,b]$ into R. Then f is continuous at $x_0 \in [a,b]$ if and only if for all $x \in {}^*[a,b]$,

$$x \approx x_0 \quad \text{implies} \quad f(x) \approx f(x_0).$$

Here we are writing f instead of $^*f$ for the corresponding map of $^*[a,b]$ into $^*R$.

PROOF. Let $A = [a,b]$, and let f be continuous at x_0. Then to every $\varepsilon \in R^+$ there corresponds a $\delta \in R^+$ such that

$$\models (\forall \mathbf{x} \in \mathbf{A})(|\mathbf{x} - \mathbf{x}_0| < \boldsymbol{\delta} \rightarrow |\mathbf{f}(\mathbf{x}) - \mathbf{f}(\mathbf{x}_0)| < \boldsymbol{\varepsilon}).$$

Using the Transfer Principle, we see that if $x \in {}^*A$ and $|x - x_0| < \delta$ then $|f(x) - f(x_0)| < \varepsilon$. Now, if $x \approx x_0$, then certainly $|x - x_0| < \delta$, since δ is standard. Thus we may conclude that

$$|f(x) - f(x_0)| < \varepsilon.$$

But since ε was any positive real number, $|f(x) - f(x_0)|$ is less than every positive real number, and hence must be infinitesimal. That is,

$$f(x) \approx f(x_0).$$

Conversely, suppose that $x \approx x_0$ implies $f(x) \approx f(x_0)$ for $x \in {}^*A$. Now choose some fixed ε in R^+. By selecting $\delta \approx 0$, we see that

$$* \models (\exists \boldsymbol{\delta} \in {}^*\mathbf{R}^+)(\forall \mathbf{x} \in {}^*\mathbf{A})(|\mathbf{x} - \mathbf{x}_0| < \boldsymbol{\delta} \rightarrow |\mathbf{f}(\mathbf{x}) - \mathbf{f}(\mathbf{x}_0)| < \boldsymbol{\varepsilon}).$$

By the Transfer Principle, f satisfies the standard condition for being continuous at x_0. ∎

Next we use this nonstandard characterization of continuity to give a nonstandard proof of a classical theorem:

THEOREM 5.5. Let f be a continuous real valued function on $[a,b]$. Let $f(a) < 0 < f(b)$. Then there exists a c such that $a < c < b$ and $f(c) = 0$.

PROOF. We begin by defining the sequence

$$t_i^{(n)} = \begin{cases} a + \dfrac{b-a}{n} i & \text{if } n \in N, 0 \leq i \leq n \\ 0 & \text{otherwise.} \end{cases} \tag{‡}$$

This is the sequence of points dividing the interval $[a,b]$ into n equal parts. We can view this sequence as a map:

$$t: N \times N \rightarrow R, \quad \text{that is} \quad t(\langle n,i \rangle) = t_i^{(n)}.$$

Then $^*t$ maps $^*N \times {}^*N$ into $^*R$. We write $t_i^{(n)}$ for $^*t \upharpoonright \langle n,i \rangle$ even for n, $i \in {}^*N$. By the Transfer Principle, (‡) holds for all $i,n \in {}^*N$.

Now choose some $\nu \in {}^*N - N$, and consider the set

$$L = \{i \in {}^*N | f(t_i^{(\nu)}) > 0, i \leq \nu\}.$$

L is a definable subset of $^*U$, for example, by the formula

$$\mathbf{i \in {}^*N \;\&\; (f \upharpoonright (t \upharpoonright \langle \nu,i \rangle)) > 0}.$$

Since $L \subseteq {}^*N$, it follows from the Internality Theorem that L is internal. Since $f(b) > 0$, $\nu \in L$. Hence $L \neq \emptyset$. Since L is internal and nonempty, it has a *least element* j (cf. Theorem 1–8.3). Thus

$$f(t_j^{(\nu)}) > 0.$$

Because $f(a) < 0$, $j \neq 0$. Hence

$$f(t_{j-1}^{(\nu)}) \leq 0.$$

Also $t_j^{(\nu)}$ is finite since $a \leq t_j^{(\nu)} \leq b$. Let $c = {}^\circ(t_j^{(\nu)})$, the standard part of $t_j^{(\nu)}$. So,

$$c \approx t_j^{(\nu)}.$$

But also,

$$t_j^{(\nu)} \approx t_{j-1}^{(\nu)},$$

since the difference between these two hyperreal numbers is $(b - a)/\nu$, an infinitesimal.

Hence by Theorem 5.4,

$$f(t_j^{(\nu)}) \approx f(c),$$

and

$$f(t_{j-1}^{(\nu)}) \approx f(c).$$

Since $f(c)$ is standard,

$$f(c) = {}^\circ f(t_j^{(\nu)}) = {}^\circ f(t_{j-1}^{(\nu)}).$$

But using the inequalities derived above and Theorem 4.1, we have

$${}^\circ f(t_j^{(\nu)}) \geq 0; \qquad {}^\circ f(t_{j-1}^{(\nu)}) \leq 0.$$

Hence $f(c) = 0$. ■

Another example is

THEOREM 5.6. Let f be a continuous real valued function on $[a,b]$. Then f has a maximum value.

PROOF. Let $t_i^{(n)}$ be as in the previous theorem. Then

$$\vDash \mathbf{(\forall n \in N)(\exists i \in N)}$$
$$\mathbf{(0 \leq i \leq n \;\&\; (\forall j \in N)(0 \leq j \leq n \rightarrow f \upharpoonright (t \upharpoonright \langle n,i \rangle) \geq f \upharpoonright (t \upharpoonright \langle n,j \rangle)))}.$$

This is because the finite set $\{f(t_0^{(n)}), \ldots, f(t_n^{(n)})\}$ of real numbers has a maximum. Using the Transfer Principle, and letting ν be a fixed infinite integer, we find that there is an i, $0 \le i \le \nu$ such that for all j for which $0 \le j \le \nu$,

$$f(t_i^{(\nu)}) \ge f(t_j^{(\nu)}).$$

Let $c = {}^\circ(t_i^{(\nu)})$, and let x be any standard point in $[a,b]$. By the Transfer Principle, there is a $j < \nu$ such that

$$t_j^{(\nu)} \le x \le t_{j+1}^{(\nu)}.$$

Since $(b - a)/\nu \approx 0$, it follows that $x = {}^\circ(t_j^{(\nu)})$. Using continuity,

$$f(c) = {}^\circ(f(t_i^{(\nu)})) \ge {}^\circ(f(t_j^{(\nu)})) = f(x).$$

Thus $f(c)$ is a maximum. ■

It is striking that these proofs remind one of the naive student's attempt to locate the zero of f in Theorem 5.5 as the "first" point where f stops being negative, or the maximum in Theorem 5.6 as the largest value of $f(x)$.

Next we have

THEOREM 5.7. f is uniformly continuous on $[a,b]$ if and only if for all $x,x' \in {}^*[a,b]$, $x \approx x'$ implies $f(x) \approx f(x')$.

PROOF. Let f be uniformly continuous on $[a,b]$. Then for any $\varepsilon \in R^+$, we can find a $\delta \in R^+$ such that

$$\models (\forall \mathbf{x} \in [\mathbf{a},\mathbf{b}])(\forall \mathbf{x}' \in [\mathbf{a},\mathbf{b}])(|\mathbf{x} - \mathbf{x}'| < \delta \rightarrow |\mathbf{f}(\mathbf{x}) - \mathbf{f}(\mathbf{x}')| < \varepsilon).$$

By the Transfer Principle, it follows that for all x,x' in ${}^*[a,b]$,

$$|x - x'| < \delta \quad \text{implies} \quad |f(x) - f(x')| < \varepsilon.$$

If, in fact, $x \approx x'$, then certainly $|x - x'| < \delta$, and hence $|f(x) - f(x')| < \varepsilon$. Since ε was any positive real number, $f(x) \approx f(x')$.

Conversely, let $f(x) \approx f(x')$ whenever $x,x' \in {}^*[a,b]$ and $x \approx x'$. Then for any $\varepsilon \in R^+$, we have, choosing for δ a positive infinitesimal,

$$* \models (\exists \delta \in {}^*\mathbf{R}^+)(\forall \mathbf{x} \in {}^*[\mathbf{a},\mathbf{b}])(\forall \mathbf{x}' \in {}^*[\mathbf{a},\mathbf{b}])(|\mathbf{x} - \mathbf{x}'| < \delta \rightarrow |\mathbf{f}(\mathbf{x}) - \mathbf{f}(\mathbf{x}')| < \varepsilon).$$

Using the Transfer Principle, we obtain the standard characterization of uniform continuity. ■

THEOREM 5.8. If f is continuous on $[a,b]$, then f is uniformly continuous on $[a,b]$.

PROOF. Suppose that $x,x' \in {}^*[a,b]$ and that $x \approx x'$. We want to show that $f(x) \approx f(x')$. Since $x,x' \in {}^*[a,b]$, we have that $a \le x,x' \le b$, and hence that x,x' are finite. Hence for some $t \in [a,b]$, $t = {}^\circ(x) = {}^\circ(x')$, that is, $x \approx t$ and

$x' \approx t$. Since f is continuous at t, $f(x) \approx f(t)$ and $f(x') \approx f(t)$. Finally, $f(x) \approx f(x')$. ■

6. PROLONGATION THEOREMS

Let $\{s_n | n \in {}^*N^+\}$ be a family of hyperreal numbers indexed by ${}^*N^+$. We speak of $\{s_n\}$ as a *sequence*, permitting a slight "abuse of language." Of course, $s \subseteq {}^*N^+ \times {}^*R$. We are interested in the case in which s is internal, or as we say, the sequence $\{s_n | n \in {}^*N^+\}$ is internal.

THEOREM 6.1. Let $\{s_n | n \in {}^*N^+\}$ be an internal sequence of hyperreal numbers. Let $|s_n| \leq M$, $M \in {}^*R$, for all $n \in N^+$. Then there is a $\nu \in {}^*N - N$ such that $|s_n| \leq M$ for all $n < \nu$.

That is, the inequality $|s_n| \leq M$ is "prolonged" from N^+ into some initial segment of the infinite integers.

PROOF. Let $A = \{n \in {}^*N^+ \| s_n| > M\}$. Since A is a definable subset of the internal set ${}^*N$, it is itself internal by Corollary 1–8.6.

Suppose first that $A = \emptyset$. Then the theorem holds, with ν being any infinite integer. Otherwise, since A is internal, by Theorem 8.3, Chapter 1, it has a least element $\nu \in {}^*N$. Thus for $n < \nu$, we have that $|s_n| \leq M$. Finally, $\nu \notin N^+$, since by hypothesis, A is disjoint from N^+. ■

The next result is used repeatedly.

THEOREM 6.2 (INFINITESIMAL PROLONGATION THEOREM). Let $\{s_n | n \in {}^*N^+\}$ be an internal sequence of hyperreal numbers, and let $s_n \approx 0$ for all $n \in N^+$. Then for some $\nu \in {}^*N - N$, $s_n \approx 0$ for all $n < \nu$.

To test the readers' understanding, we begin with an incorrect proof, modeled on the correct proof of the previous theorem.

FALLACIOUS PROOF. Let $A = \{n \in {}^*N | s_n \approx 0\}$. If $A = \emptyset$, we are done; otherwise let ν be the least element of A. Then for $n < \nu$, we have $s_n \approx 0$. ν must be infinite by hypothesis. ■

What is wrong? The set A need not be internal; in fact, its defining condition uses the external set I of infinitesimals. Hence A need not have a least element.

To move towards a correct proof, let us write

$$s = \{s_n | n \in {}^*N^+\} = \{\langle n, s_n \rangle | n \in {}^*N^+\}.$$

Let $t_n = n \cdot s_n$ for $n \in {}^*N^+$. Then, since s is an internal function for each

$n \in {}^*N$, we have

$$t_n = |\mathbf{n} \cdot (\mathbf{s} \restriction \mathbf{n})|_*.$$

So, by Theorem 8.9 of Chapter 1, $\{t_n | n \in {}^*N^+\}$ is also internal.

Now, we give a

CORRECT PROOF. Let $t_n = n \cdot s_n$. Since $s_n \approx 0$ for $n \in N^+$, it follows that $t_n \approx 0$ for $n \in N^+$. Hence

$$|t_n| \leq 1 \quad \text{for} \quad n \in N^+.$$

(We have traded in the "external condition" $t_n \approx 0$ for the weaker "internal condition" $|t_n| \leq 1$.) Since, as we have just seen, $\{t_n | n \in {}^*N^+\}$ is internal, Theorem 6.1 shows that for some $v \in {}^*N - N$,

$$|t_n| \leq 1 \quad \text{for} \quad n < v.$$

Then $|s_n| \leq 1/n \approx 0$ if n is infinite and $< v$. ■

As an application of the Infinitesimal Prolongation Theorem, we give a nonstandard proof of the fact that a uniformly convergent sequence of continuous functions has a continuous limit. First we must pause to derive the appropriate nonstandard equivalent of uniform convergence. We write $s_n(x) \rightrightarrows s(x)$ to mean that $s_n(x)$ converges uniformly to $s(x)$.

THEOREM 6.3. $s_n(x) \rightrightarrows s(x)$ on $[a,b]$ if and only if $s_v(x) \approx s(x)$ for $x \in {}^*[a,b]$ and $v \in {}^*N - N$.

Note that as with uniform continuity, the passage from a "pointwise" to a "uniform" notion, takes the form of the nonstandard condition holding on all of ${}^*[a,b]$ instead of just $[a,b]$.

PROOF. Let $s_n(x) \rightrightarrows s(x)$ on $[a,b]$. Then for any real $\varepsilon > 0$, there exists an $n_0 \in N$ such that

$$\models (\forall \mathbf{n} \in \mathbf{N})(\forall \mathbf{x} \in [\mathbf{a},\mathbf{b}])(\mathbf{n} > \mathbf{n}_0 \rightarrow |\mathbf{s}_n(\mathbf{x}) - \mathbf{s}(\mathbf{x})| < \varepsilon).$$

By the Transfer Principle, for all $n \in {}^*N$,

$$n > n_0 \text{ and } x \in {}^*[a,b] \text{ implies } |s_n(x) - s(x)| < \varepsilon.$$

Now if $x \in {}^*[a,b]$ and $v \in {}^*N - N$, then $v > n_0$, and so

$$|s_v(x) - s(x)| < \varepsilon.$$

Since this holds for all $\varepsilon \in R^+$, we have $s_v(x) \approx s(x)$.

Conversely, let ε be any positive real number. Choosing $n_0 \in {}^*N - N$, we have

$$* \models (\exists \mathbf{n}_0 \in {}^*\mathbf{N})(\forall \mathbf{n} \in {}^*\mathbf{N})(\forall \mathbf{x} \in {}^*[\mathbf{a},\mathbf{b}])(\mathbf{n} > \mathbf{n}_0 \rightarrow |\mathbf{s}_n(\mathbf{x}) - \mathbf{s}(\mathbf{x})| < \varepsilon)).$$

Our result follows in the usual way applying the Transfer Principle. ■

THEOREM 6.4. Let $s_n(x)$ be continuous on $[a,b]$ for each $n \in N^+$. Let $s_n(x) \rightrightarrows s(x)$ on $[a,b]$. Then $s(x)$ is continuous on $[a,b]$.

PROOF. Let $x_0 \in [a,b]$ and $x \in {}^*[a,b]$ where $x \approx x_0$. We want to show that $s(x) \approx s(x_0)$. Now, for each $n \in N^+$, we know that $s_n(x) \approx s_n(x_0)$, that is, $s_n(x) - s_n(x_0) \approx 0$. So by the Infinitesimal Prolongation Theorem, $s_\nu(x) - s_\nu(x_0) \approx 0$ for some $\nu \in {}^*N - N$. Using Theorem 6.3,

$$s(x) \approx s_\nu(x) \approx s_\nu(x_0) \approx s(x_0). \quad \blacksquare$$

For another example of the prolongation technique, we consider the nonstandard version of Cauchy's convergence criterion:

THEOREM 6.5. Let $\{s_n | n \in N^+\}$ be a sequence of real numbers. Let $s_n \approx s_m$ for all $n,m \in {}^*N - N$. Then $s_n \to L$ for some real number L.

PROOF. It suffices to show that for all $n \in {}^*N - N$, s_n is finite. For choose some definite $\nu \in {}^*N - N$. Then for $L = {}^\circ(s_\nu)$, we would have $s_\nu \approx L$. From the hypothesis, it would then follow that for all $\mu \in {}^*N - N$, $s_\mu \approx s_\nu \approx L$, that is, by Theorem 5.1 $s_n \to L$.

Thus let $\nu \in {}^*N - N$. We wish to show that s_ν is finite. Consider the set

$$A = \{n \in {}^*N^+ \,|\, |s_n - s_\nu| < 1\}.$$

Since A is definable in ${}^*U$, and $A \subseteq {}^*N$, A is internal. Now for $n \in {}^*N - N$, $s_n \approx s_\nu$, and, therefore, $|s_n - s_\nu| < 1$; thus ${}^*N - N \subseteq A$. Since A is internal and ${}^*N - N$ is external (Corollary 1–8.4), ${}^*N - N \neq A$. Hence there exists an $n_0 \in A \cap N^+$. Then,

$$|s_\nu| = |s_{n_0} - (s_{n_0} - s_\nu)| \leq |s_{n_0}| + |s_{n_0} - s_\nu| < |s_{n_0}| + 1,$$

so that s_ν is finite. $\blacksquare$

7. NONSTANDARD DIFFERENTIAL CALCULUS

In this section, we briefly develop the calculus for functions of one variable in the spirit of Leibniz. Our development differs slightly from those in the literature and is intended to generalize readily to yield the Fréchet derivative (cf. Section 4 of Chapter 4).

We recall that I is the set of infinitesimals of ${}^*R$.

DEFINITION. Let f,g each map I into I. We write $f \sim g$ and call f and g *equivalent* if for all nonzero $h \approx 0$,

$$\frac{f(h) - g(h)}{h} \approx 0.$$

Writing $(f(h) - g(h))/h = \alpha$, this condition can be equivalently written as

$$f(h) = g(h) + \alpha \cdot h, \qquad \alpha \approx 0.$$

Here $\alpha \cdot h$ is an infinitesimal of "higher order" than h in the sense that $(\alpha \cdot h)/h$ is infinitesimal. So, $f \sim g$ simply means that $f(h) - g(h)$ is an infinitesimal of higher order than h.

DEFINITION. f is called *locally linear* if f maps I into I and, for all $\alpha,\beta \approx 0$,

$$f(\alpha\beta) = \alpha f(\beta).$$

For a locally linear function, we have

$$f(0) = f(\alpha \cdot 0) = \alpha f(0)$$

for any $\alpha \approx 0$. Hence $f(0) = 0$.

THEOREM 7.1. Let f be locally linear. Then there is a finite $k \in {}^*R$ such that $f(\alpha) = k\alpha$ for all $\alpha \approx 0$.

PROOF. Let $\alpha,\beta \approx 0$, $\alpha,\beta \neq 0$. Then

$$\frac{f(\beta)}{\beta} = \frac{\alpha f(\beta)}{\alpha\beta} = \frac{f(\alpha\beta)}{\alpha\beta} = \frac{\beta f(\alpha)}{\alpha\beta} = \frac{f(\alpha)}{\alpha}.$$

Let k be the constant value of $f(\alpha)/\alpha$ for $\alpha \approx 0$, $\alpha \neq 0$. Then $f(\alpha) = k\alpha$ for all $\alpha \approx 0$. (The equation holds automatically for $\alpha = 0$, since $f(0) = 0$.)

To see that k is finite, assume the contrary. Then $1/k \approx 0$, and

$$f\left(\frac{1}{k}\right) = k \cdot \frac{1}{k} = 1 \not\approx 0,$$

which contradicts the fact that f maps I into I. ■

THEOREM 7.2. Let $f_1(\alpha) = k_1\alpha$, $f_2(\alpha) = k_2\alpha$ be locally linear. Then

$$f_1 \sim f_2 \quad \text{if and only if} \quad k_1 \approx k_2.$$

PROOF. We have

$$\frac{f_1(\alpha) - f_2(\alpha)}{\alpha} = \frac{k_1\alpha - k_2\alpha}{\alpha} = k_1 - k_2,$$

so that the left side is infinitesimal if and only if the right side is. ■

DEFINITION. If $f(\alpha) = k\alpha$ where $k \in R$, then f is called a *differential.*

COROLLARY 7.3. Every locally linear function is equivalent to a unique differential.

PROOF. The proof is immediate from Theorems 7.1 and 7.2. ■

Following Leibniz, we employ the symbols dx,dt,dy as variables whose values are infinitesimals.

DEFINITION. Let f be a real function defined at x_0. The function $\Delta_{x_0}f$ is defined by

$$\Delta_{x_0}f(dx) = f(x_0 + dx) - f(x_0).$$

Note that f is continuous at x_0 if and only if $\Delta_{x_0}f$ maps I into I.

DEFINITION. f is *differentiable* at x_0 if $\Delta_{x_0}f$ is equivalent to some differential.

By what we have shown, if f is differentiable at x_0, then $\Delta_{x_0}f$ is equivalent to a *unique* differential, written $d_{x_0}f$ or just df, and called the *differential of* f (at x_0). In this case, we have for all $dx \in I$,

$$d_{x_0}f(dx) = k\,dx$$

for some real number k. This number k is called the derivative of f at x_0, writen $f'(x_0)$. We usually write (to conform with Leibniz' notation) $d_{x_0}f$ or just $d_{x_0}f(dx)$. Thus we have

$$f'(x_0) = \frac{df}{dx}, \tag{‡}$$

and the derivative is truly the quotient of two infinitesimals.[1]

Note that to say that f is differentiable at x_0 with derivative $f'(x_0)$ is just to say that $\Delta_{x_0}f(dx) \sim f'(x_0)\,dx$, that is, that

$$\frac{f(x_0 + dx) - f(x_0) - f'(x_0)\,dx}{dx} = \frac{f(x_0 + dx) - f(x_0)}{dx} - f'(x_0) \approx 0,$$

for all $dx \approx 0$, that is, that

$$\lim_{h\to 0} \frac{f(x_0 + h) - f(x_0)}{h} = f'(x_0).$$

This, of course, is the standard definition of derivative.

THEOREM 7.4. If f is differentiable at x_0, then it is continuous at x_0.

PROOF. Let $\Delta_{x_0}f(dx) \sim k\,dx$. That is,

$$f(x_0 + dx) - f(x_0) = k\,dx + \alpha\,dx$$

[1] In the treatment of Robinson [14], df is defined to be what we've called Δf. Thus (‡) does not in general hold in Robinson's treatment. Instead, one has, $f'(x_0) = {}^{\circ}(df/dx)$.

where $\alpha \approx 0$. Then $f(x_0 + dx) \approx f(x_0)$ for $dx \approx 0$, that is, f is continuous at x_0. ■

As an example, let $f(x) = x^2$. Then,

$$\begin{aligned}\Delta_{x_0} f &= (x_0 + dx)^2 - x_0^2 \\ &= 2x_0\, dx + dx^2.\end{aligned}$$

Since dx^2 is an infinitesimal of higher order than dx, $(dx^2/dx = dx \approx 0)$, $\Delta_{x_0} f \sim 2x_0\, dx$. Hence $df = 2x\, dx$, $df/dx = 2x$.

The condition $\Delta_{x_0} f \sim f'(x_0)\, dx$ can be written equivalently:

$$f(x_0 + dx) - f(x_0) = f'(x_0)\, dx + \alpha\, dx, \qquad \alpha \approx 0,$$

that is,

$$f(x_0 + dx) = f(x_0) + f'(x_0)\, dx + \alpha\, dx, \qquad \alpha \approx 0.$$

THEOREM 7.5. If f,g are differentiable at x_0, so is fg, and

$$d(fg) = f\, dg + g\, df.$$

PROOF.
$$\begin{aligned}f(x_0 + dx) &= f(x_0) + d_{x_0} f + \alpha\, dx, \\ g(x_0 + dx) &= g(x_0) + d_{x_0} g + \beta\, dx,\end{aligned}$$

$\alpha,\beta \approx 0$. Thus

$$\begin{aligned}f(x_0 + dx)g(x_0 + dx) = {} & f(x_0)g(x_0) + f\, dg + g\, df \\ & + \{df \cdot dg + \beta f\, dx + \beta\, df \cdot dx \\ & + \alpha\beta\, dx^2 + \alpha g\, dx + \alpha\, dg \cdot dx\}.\end{aligned}$$

To obtain the result, we need only check that each term in the curly braces is an infinitesimal of higher order than dx. This can be seen by inspection for all of them except $df \cdot dg$. But

$$df \cdot dg = df \cdot g'(x_0)\, dx,$$

which is of higher order than dx because $df \approx 0$. ■

THEOREM 7.6 (CHAIN RULE). Let $k(x) = f(g(x))$, where g is differentiable at x_0 and f is differentiable at $g(x_0)$. Then k is differentiable at x_0 and $dk = f'(g(x_0))\, dg$.

PROOF. Let $u_0 = g(x_0)$. Then for some $\alpha \approx 0$,

$$g(x_0 + dx) = u_0 + g'(x_0)\, dx + \alpha\, dx.$$

Writing $du = g'(x_0)\, dx + \alpha\, dx \approx 0$, we have

$$\begin{aligned}f(g(x_0 + dx)) &= f(u_0 + du) \\ &= f(u_0) + f'(u_0)\, du + \beta\, du\end{aligned}$$

for some $\beta \approx 0$. Thus

$$\Delta_{x_0} k = f'(g(x_0))g'(x_0)\,dx + \{f'(u_0)\alpha\,dx + \beta\,du\},$$

which gives the result. ■

8. ADDITIVITY[1]

In this section, we consider three miscellaneous applications of nonstandard methods that involve additivity.

We first consider the functional equation:

$$f(x + y) = f(x) + f(y). \qquad (\ddagger)$$

It is an easy result due to Cauchy that the only *continuous* functions f mapping R into R that satisfy (‡) are the functions $f(x) = kx$ for some $k \in R$. Many generalizations of Cauchy's result have been obtained. We give a nonstandard proof of the following:

THEOREM 8.1 (DARBOUX). Let f map R into R satisfying (‡), and let f be bounded on some interval. Then $f(x) = f(1) \cdot x$.

PROOF. We begin with some easy consequences of (‡).

(1) $f(0) = 0$. For, $f(0) = f(0) + f(0)$.
(2) *For all* $n \in N$, $f(nx) = nf(x)$. By induction on n. For $n = 0$, the result is (1). Assume the result holds for n. Then

$$\begin{aligned} f((n+1)x) &= f(nx + x) \\ &= f(nx) + f(x) \\ &= nf(x) + f(x) \\ &= (n+1)f(x). \end{aligned}$$

(3) *For all* $x \in R$, $f(-x) = -f(x)$, since $f(x) + f(-x) = f(0) = 0$.
(4) *For all integers* n, $f(nx) = nf(x)$. By (2), the result holds for $n \geq 0$. Let $n < 0$. Then

$$\begin{aligned} f(nx) &= f(-(-n)x) \\ &= -f((-n)x) \\ &= -(-n)f(x) \\ &= nf(x). \end{aligned}$$

[1] This and the following section are largely based on Luxemburg [8, 11].

(5) *For any rational* q, $f(qx) = qf(x)$. Let $q = m/n$, m,n integers, $n \neq 0$. Then

$$nf\left(\frac{m}{n}x\right) = f(mx)$$
$$= mf(x).$$

Thus

$$f\left(\frac{m}{n}x\right) = \frac{m}{n}f(x).$$

For rational q, (5) yields $f(q) = f(1) \cdot q$, and the desired result could be obtained at once *for continuous* f using the density of rational numbers in R.

Now, let I be an interval on which f is bounded, and let x_0 be some interior point of I. Let $|f(x)| \leq M$, $M \in R$, for all $x \in I$. We proceed by first showing that $h \approx 0$ implies $f(h) \approx 0$. Since x_0 is an interior point of I, there is a real $\delta > 0$ such that

$$|f(x_0 + h)| \leq M$$

for all $|h| < \delta$, hence (by the Transfer Principle) for all $h \approx 0$.

Hence for $h \approx 0$,

$$\begin{aligned} |f(h)| &= |f(x_0 + h) - f(x_0)| \\ &\leq |f(x_0 + h)| + |f(x_0)| \\ &\leq M + |f(x_0)|. \end{aligned}$$

Now, $h \approx 0$ and $n \in N$ implies $nh \approx 0$. Hence $n|f(h)| = |f(nh)| \leq M + |f(x_0)|$. That is,

$$|f(h)| \leq \frac{M + |f(x_0)|}{n}$$

for all $n \in N^+$. Hence $f(h) \approx 0$.

Now let $x \in R$. By Theorem 4.5, $x = q + h$, where $q \in {}^*Q$ and $h \approx 0$. Then (using the Transfer Principle twice),

$$f(x) = f(q) + f(h) \approx f(q) = qf(1) \approx xf(1).$$

Hence $f(x) = xf(1)$. ■

Our next example is a generalization by Luxemburg of a result from Pólya-Szegö's collection of problems. (The case $p = 0$ is the Pólya-Szegö result.)

THEOREM 8.2. Let $\{s_n | n \in N^+\}$ be a sequence of real numbers such that $|s_{n+m} - s_n - s_m| \leq s(n^p + m^p)$ for all $n,m \in N^+$, some p, $0 \leq p < 1$, and some $s \in R$. Then, there is a $\sigma \in R$ such that

(1) $\dfrac{s_n}{n} \to \sigma$,

(2) $|s_n - n\sigma| \leq s\left(\dfrac{n^p}{1 - 2^{p-1}}\right).$

We remark that (1) follows immediately from (2), since $p < 1$ implies $n^{p-1}/(1 - 2^{p-1}) \to 0$.

The hypothesis of the theorem is a kind of approximate additivity. Iterating it, we obtain the

LEMMA. Under the hypothesis of the theorem,

$$\left|\frac{s_{2^k n}}{2^k} - s_n\right| \leq sn^p \sum_{i=0}^{k-1} 2^{i(p-1)} = sn^p\left(\frac{1 - 2^{(p-1)k}}{1 - 2^{p-1}}\right).$$

PROOF. The proof is by induction on k. By hypothesis (setting $m = n$),

$$|s_{2n} - 2s_n| \leq 2sn^p.$$

But this is the case $k = 1$ of the lemma. Assuming the result known for some given k, we have

$$\left|\frac{s_{2^{k+1}n}}{2^{k+1}} - s_n\right| \leq \left|\frac{s_{2^{k+1}n}}{2^{k+1}} - \frac{s_{2n}}{2}\right| + \left|\frac{s_{2n}}{2} - s_n\right|.$$

The second term on the right is $\leq sn^p$ by the case $k = 1$. By the induction hypothesis,

$$\left|\frac{s_{2^k(2n)}}{2^k} - s_{2n}\right| \leq s(2n)^p \sum_{i=0}^{k-1} 2^{i(p-1)}.$$

Thus

$$\left|\frac{s_{2^{k+1}n}}{2^{k+1}} - s_n\right| \leq sn^p\left\{1 + \sum_{i=0}^{k-1} 2^{(i+1)(p-1)}\right\} = sn^p \sum_{i=0}^{k} 2^{i(p-1)}. \quad ■$$

PROOF OF THEOREM. In the lemma, let $k = \nu \in {}^*N - N$. Thus

$$\left|\frac{s_{2^\nu n}}{2^\nu} - s_n\right| \leq sn^p\left(\frac{1 - 2^{(p-1)\nu}}{1 - 2^{p-1}}\right) \approx \frac{sn^p}{1 - 2^{p-1}}. \qquad (*)$$

Hence $\dfrac{s_{2^\nu n}}{2^\nu}$ is finite for all $n \in N^+$. Let $t_n = {}^\circ\!\left(\dfrac{s_{2^\nu n}}{2^\nu}\right)$. By hypothesis,

$$|s_{2^\nu(n+m)} - s_{2^\nu n} - s_{2^\nu m}| \leq s \cdot 2^{\nu p}(n^p + m^p).$$

Hence

$$\left|\frac{s_{2^\nu(n+m)}}{2^\nu} - \frac{s_{2^\nu n}}{2^\nu} - \frac{s_{2^\nu m}}{2^\nu}\right| \leq s \cdot 2^{\nu(p-1)}(n^p + m^p).$$

Since $p - 1 < 0$, the right side of the inequality is infinitesimal, and, taking standard parts, $|t_{n+m} - t_n - t_m| = 0$. That is, $t_{n+m} = t_n + t_m$. But (as, for example, we saw in the proof of Theorem 7.1), this implies

$$t_n = nt_1 = n \cdot \sigma,$$

where $\sigma = {}^{\circ}\left(\frac{s_{2^\nu}}{2^\nu}\right)$. Recalling the inequality (*) and taking standard parts, we see that

$$|t_n - s_n| = |n\sigma - s_n| \leq \frac{sn^p}{1 - 2^{p-1}},$$

the desired result. ■

Our final example is a nonstandard proof of the existence of so-called Banach limits.

DEFINITION. A map LIM of the set of bounded sequences of real numbers into R is called a *Banach limit* if it satisfies the axioms:

(1) $s_n \geq 0$ implies $\text{LIM}(s_n) \geq 0$.
(2) $\text{LIM}(\alpha s_n + \beta t_n) = \alpha\ \text{LIM}(s_n) + \beta\ \text{LIM}(t_n)$ for $\alpha, \beta \in R$.
(3) $\liminf s_n \leq \text{LIM}(s_n) \leq \limsup s_n$.
(4) $\text{LIM}(s_{n+1}) = \text{LIM}(s_n)$.

We prove

THEOREM 8.3. There exists a Banach limit.

PROOF. To begin with consider the average

$$a_k = \frac{1}{k} \sum_{n=k}^{2k-1} s_n.$$

Then this equation must hold for all $k \in {}^*N^+$. For k finite, there is some k_0 such that $|a_k| \leq |s_{k_0}| = \max_{k \leq n < 2k} |s_n|$. Hence by the Transfer Principle, the same is true when k is infinite. If $\{s_n | n \in N^+\}$ is a bounded sequence $|s_n| \leq M$, then (using the Transfer Principle again) each s_n for $n \in {}^*N$ must be finite. It follows that for a bounded sequence $\{s_n\}$, a_k is finite for all $k \in {}^*N^+$. So we choose some fixed $\nu \in {}^*N - N$ and define

$$\text{LIM}(s_n) = {}^{\circ}(a_\nu).$$

We now proceed to verify Axioms 1–4.

(1) If $s_n \geq 0$ for all $n \in N^+$, then $a_n \geq 0$ for all $n \in N^+$. By the Transfer Principle, $a_\nu \geq 0$, so $\text{LIM}(s_n) \geq 0$.

(2) Let $u_n = \alpha s_n + \beta t_n$. Then

$$\begin{aligned}\mathrm{LIM}(u_n) &= {}^\circ\left(\frac{1}{\nu}\sum_{n=\nu}^{2\nu-1} u_n\right) = {}^\circ\left(\frac{1}{\nu}\sum_{n=\nu}^{2\nu-1} (\alpha s_n + \beta t_n)\right)\\ &= {}^\circ\left(\alpha\frac{1}{\nu}\sum_{n=\nu}^{2\nu-1} s_n + \beta\frac{1}{\nu}\sum_{n=\nu}^{2\nu-1} t_n\right)\\ &= \alpha\,{}^\circ\left(\frac{1}{\nu}\sum_{n=\nu}^{2\nu-1} s_n\right) + \beta\,{}^\circ\left(\frac{1}{\nu}\sum_{n=\nu}^{2\nu-1} t_n\right)\\ &= \alpha\,\mathrm{LIM}(s_n) + \beta\,\mathrm{LIM}(t_n).\end{aligned}$$

(3) Let ε be positive and real. Then (cf. Theorem 5.3) ${}^\circ(s_n)$ is a limit point of the sequence $\{s_n | n \in N^+\}$ for each $n \in {}^*N - N$. Hence for each such n,

$$\liminf_{n\to\infty} \{s_n\} \le {}^\circ(s_n) \le \limsup_{n\to\infty} \{s_n\},$$

and hence

$$\liminf_{n\to\infty} \{s_n\} - \varepsilon \le s_n \le \limsup_{n\to\infty} \{s_n\} + \varepsilon.$$

Since this holds in particular, for all n such that $\nu \le n < 2\nu$, we obtain

$$\liminf_{n\to\infty} \{s_n\} - \varepsilon \le \frac{1}{\nu}\sum_{n=\nu}^{2\nu-1} s_n \le \limsup_{n\to\infty} \{s_n\} + \varepsilon.$$

Taking standard parts,

$$\liminf_{n\to\infty} \{s_n\} - \varepsilon \le \mathrm{LIM}(s_n) \le \limsup_{n\to\infty} \{s_n\} + \varepsilon.$$

The result follows by letting $\varepsilon \to 0$.

(4) Let $u_n = s_{n+1}$. Then

$$\begin{aligned}\mathrm{LIM}(u_n) &= {}^\circ\left(\frac{1}{\nu}\sum_{n=\nu}^{2\nu-1} s_{n+1}\right)\\ &= {}^\circ\left(\frac{1}{\nu}\left\{\sum_{n=\nu}^{2\nu-1} s_n - s_{2\nu} + s_\nu\right\}\right)\\ &= {}^\circ\left(\frac{1}{\nu}\sum_{n=\nu}^{2\nu-1} s_n\right) - {}^\circ\left(\frac{s_{2\nu}}{\nu}\right) + {}^\circ\left(\frac{s_\nu}{\nu}\right)\\ &= \mathrm{LIM}(s_n) - 0 + 0 = \mathrm{LIM}(s_n). \quad ■\end{aligned}$$

9. THE EXISTENCE OF NONMEASURABLE SETS

In this section, we present a curious and rather untypical use of nonstandard analysis to prove the existence of a bounded set of real numbers that is not

Lebesgue measurable. We assume (in this section only) that the reader is familiar with Lebesgue measure and integration.

As usual, a function f that maps R into R is said to have p as a *period* if $f(x + p) = f(x)$ for all $x \in R$. We recall that if p is a period of a bounded measurable function f, then

$$\int_x^{x+p} f(t)\,dt = \int_y^{y+p} f(t)\,dt$$

for any $x,y \in R$. We shall make use of the following:

THEOREM 9.1. Let f be a bounded measurable function on R with arbitrarily small periods. Then f is constant almost everywhere (a.e.).

PROOF. By the Fundamental Theorem of Calculus for the Lebesgue integral, we have for any $x,y \in R$:

$$f(x) \underset{\text{a.e.}}{=} \lim_{h\to 0} \frac{1}{h} \int_x^{x+h} f(t)\,dt = \lim_{h\to 0} \frac{1}{h} \int_y^{y+h} f(t)\,dt \underset{\text{a.e.}}{=} f(y),$$

where the values of h are restricted to periods of f (as they may be by hypothesis). Hence $f(x) = f(y)$ a.e., that is, $f(x)$ is constant almost everywhere.

DEFINITION. A *binary rational number* is one of the form $\sum_{k=1}^{n} a_k/2^k$, where each a_k is 0 or 1. We write B for the set of binary rationals.

Of course, each binary rational number r is a rational number such that $0 \le r < 1$.

For each *real* number x, $0 \le x < 1$, let $s_x(n)$ be the nth "digit" in the binary expansion of x. That is,

$$x = \sum_{n=1}^{\infty} \frac{s_x(n)}{2^n},$$

where each $s_x(n) = 0$ or 1. (For *binary rationals* r this description does not uniquely determine $s_r(n)$; we eliminate this ambiguity by insisting that $s_r(n) = 0$ for all sufficiently large n). For each x, s_x is a map of N^+ into $\{0,1\}$; hence for each x, $^*(s_x)$ is a map of $^*N^+$ into $\{0,1\}$. (Since $\{0,1\}$ is finite, $^*\{0,1\} = \{0,1\}$.) As usual, we write s_x instead of $^*(s_x)$.

We choose some fixed $v \in {^*N} - N$ and let A be the set of real x, $0 \le x < 1$, for which $s_x(v) = 1$. We shall show that *A is not Lebesgue measurable*. Hence we suppose that A is Lebesgue measurable with measure $m(A)$,

$$0 \le m(A) \le 1,$$

and we proceed to derive a contradiction.

Each real number x can be uniquely written in the form $m + y$, where m is an integer and $0 \leq y < 1$. Then we define

$$f(x) = \begin{cases} 1 & \text{if } y \in A \\ 0 & \text{if } y \notin A \end{cases}$$

Since we are assuming that A is measurable, it follows that f is a bounded measurable function on R.

LEMMA 1. $r \in B$ implies $f(x + r) = f(x)$ for all $x \in R$.

PROOF. We can assume that $r \neq 0$. ■

CASE 1

$0 \leq x, x + r < 1$. We have to show that $x \in A$ if and only if $x + r \in A$, that is, that $s_x(v) = s_{x+r}(v)$. To do this we observe that for some finite integer m,

$$\models (\forall \mathbf{n} \in \mathbf{N}^+)(\mathbf{n} > \mathbf{m} \to \mathbf{s}_{x+r}(\mathbf{n}) = \mathbf{s}_x(\mathbf{n}));$$

this is because adding r to x alters only finitely many digits in the binary expansion of x. Using the Transfer Principle, we conclude that

$$s_{x+r}(v) = s_x(v).$$

CASE 2

$m \leq x, x + r < m + 1$, *where m is an integer*. Setting $x' = x - m$, we have $0 \leq x', x' + r < 1$, and Case 1 applies. We then conclude that

$$f(x + r) = f(x' + r) = f(x') = f(x).$$

CASE 3

$x < m \leq x + r$, *where m is an integer*. Let $x' = x - m + r$. Then $x' = (x + r) - m \geq 0$; $x' = x - m + r < r < 1$; $x' + (1 - r) > x' \geq 0$; $x' + (1 - r) = x - (m - 1) < m - (m - 1) = 1$. That is,

$$0 \leq x', \qquad x' + (1 - r) < 1,$$

where, of course, $1 - r \in B$. Then, using Case 1,

$$f(x + r) = f(x') = f(x' + (1 - r)) = f(x' - r) = f(x). \quad ■$$

Since B contains arbitrarily small elements, we conclude from Theorem 8.1 that f is constant almost everywhere. That is, either $f(x) = 0$ a.e. or $f(x) = 1$ a.e. Hence we have

LEMMA 2. $m(A) = 0$ or $m(A) = 1$.

LEMMA 3. If $0 < x < 1$ and $x \notin B$, then $s_{1-x}(n) = 1 - s_x(n)$ for $\mathrm{n} \in \mathrm{N}^+$.

PROOF.

$$\sum_{n=1}^{\infty} \frac{s_x(n)}{2^n} + \sum_{n=1}^{\infty} \frac{1 - s_x(n)}{2^n} = \sum_{n=1}^{\infty} \frac{1}{2^n} = 1.$$

Hence

$$\sum_{n=1}^{\infty} \frac{1 - s_x(n)}{2^n} = 1 - x. \quad \blacksquare$$

Using Lemma 3 and the Transfer Principle, we conclude that for $x \notin B$, $s_{1-x}(v) = 1 - s_x(v)$. Hence $x \in A - B$ if and only if $1 - x \in \bar{A} - B$, where where $\bar{A} = \{x \in R | 0 \leq x < 1, x \notin A\}$.

Since the transformation $x \to 1 - x$ is simply a Euclidean motion (i.e., a reflection about the point $x = \frac{1}{2}$), it preserves measure. Thus

$$m(A - B) = m(\bar{A} - B).$$

Since B is countable, with measure 0,

$$m(A - B) = m(A)$$

and

$$m(\bar{A} - B) = m(\bar{A}).$$

Hence we conclude that

$$m(A) = m(\bar{A}) = \tfrac{1}{2}.$$

But this contradicts Lemma 2.

EXERCISES

1. Prove that there is an element $x \in {}^*R - R$ such that $x \notin {}^*Q$.
2. Show that Theorem 4.5 holds for x hyperreal.
3. (a) Show that a sequence $\{s_n | n \in N^+\}$ of real numbers is bounded if and only if s_n is finite for all infinite n.
   (b) Now use Theorem 5.3 to show that every bounded sequence of real numbers has a limit point.
4. (a) Let f map $[a,b]$ into R, $a < b$. Let $r = \{\langle x,y \rangle | a \leq x,y \leq b$ and $f(x) \leq f(y)\}$. Show that r is concurrent.
   (b) Use (a) to give another proof of Theorem 5.6.
5. Give a nonstandard proof that $s_n \to L$ implies

$$(s_1 + s_2 + \cdots + s_n)/n \to L$$

   (Hint: Without loss of generality, assume $L = 0$, $s_n \geq 0$. For $v \in {}^*N - N$, write $s_1 + s_2 + \cdots + s_v = (s_1 + \cdots + s_\mu) + (s_{\mu+1} + \cdots + s_v)$ where $\mu/v \approx 0$.)
6. Show that for any $x \in {}^*R$, $\{y \in {}^*R | y \approx x\}$ is external.

3

TOPOLOGICAL AND METRIC SPACES

1. TOPOLOGICAL SPACES

In order to motivate our nonstandard treatment of topological spaces, we reexamine the topology of the real number system viewed from the perspective of the hyperreal numbers. Topological notions such as limit and continuity were systematically reduced to the relation $x \approx x_0$, where $x_0 \in R$ and $x \in {}^*R$. This suggests that we think of each real point x_0 as being included in an "infinitesimal neighborhood" consisting of all points $x_0 + h$, where $h \approx 0$. We shall write this set $\mu(x_0)$, called (after Leibniz) the *monad* of x_0. That is,

$$\mu(x_0) = \{x_0 + h | h \approx 0\}.$$

To see how this notion might generalize to a more abstract setting, for each $r \in R^+$, let

$$I_r = \{x \in R | x_0 - r < x < x_0 + r\},$$

and note that

$$\mu(x_0) = \bigcap_{r \in R^+} {}^*I_r.$$

With this hint, we proceed with the general case. We recall the

DEFINITION. A *topological space* is a set X together with a collection $\mathcal{O} \subseteq P(X)$ such that

(1) $\varnothing, X \in \mathcal{O}$,
(2) $G_1, \ldots, G_n \in \mathcal{O}$ implies $G_1 \cap \cdots \cap G_n \in \mathcal{O}$,
(3) If $G_i \in \mathcal{O}$ for each $i \in J$, then $\bigcup_{i \in J} G_i \in \mathcal{O}$.

The elements of $\mathcal{O}$ are then called *open sets*.

In order to use nonstandard methods, we assume that $X \in U$. (Note that we are not assuming that $X \subseteq S$.) Then each subset of X belongs to U as does $\mathcal{O}$.

For each $p \in X$, we write

$$\mathcal{O}_p = \{G \in \mathcal{O} | p \in G\}.$$

We refer to an element of $\mathcal{O}_p$ as an *open neighborhood* of p.

DEFINITION. $\mu(p) = \bigcap\{{}^*G | G \in \mathcal{O}_p\}$. We call $\mu(p)$ the *monad* of p. We write $q \approx p$ for $q \in \mu(p)$.

Note that this relation $q \approx p$ can only hold for $q \in {}^*X$ and $p \in X$. In particular, it is not ordinarily symmetric.

As the example of $^*R$ suggests, monads need not be internal sets. However, each monad contains an internal subset which behaves like an open neigborhood. More precisely,

THEOREM 1.1. For each $p \in X$ there is an internal set $D \in {}^*\mathcal{O}_p$ such that $D \subseteq \mu(p)$.

PROOF. Let $r = \{\langle x,y \rangle | x \in \mathcal{O}_p,\ y \in \mathcal{O}_p,\ y \subseteq x\}$. We show that r is concurrent.

Clearly, $\text{dom}(r) = \mathcal{O}_p$. To see that r is concurrent, let $A_1, \ldots, A_k \in \mathcal{O}_p$. Then $B = \bigcap_{i=1}^{k} A_i \in \mathcal{O}_p$, so that $\langle A_i, B \rangle \in r$, $i = 1,2, \ldots, k$, and hence r is concurrent.

By the Concurrence Theorem, there exists a $D \in {}^*U$ such that for all $A \in \mathcal{O}_p$, $\langle {}^*A, D \rangle \in {}^*r$. We have

$$\models (\forall \mathbf{x} \in \mathcal{O}_p)(\forall \mathbf{y} \in \mathcal{O}_p)(\langle \mathbf{x},\mathbf{y} \rangle \in \mathbf{r} \rightarrow \mathbf{y} \subseteq \mathbf{x}).$$

Since ${}^*r \subseteq {}^*\mathcal{O}_p \times {}^*\mathcal{O}_p$, the Transfer Principle yields $D \subseteq {}^*A$. Hence

$$D \subseteq \bigcap\{{}^*A | A \in \mathcal{O}_p\} = \mu(p). \quad \blacksquare$$

Recall that a topological space X is called a *Hausdorff space* (or a T_2-*space*) if any two distinct points are contained in disjoint open sets.

THEOREM 1.2. X is Hausdorff if and only if $p,q \in X$, and $p \neq q$ implies $\mu(p) \cap \mu(q) = \varnothing$.

PROOF. Let X be Hausdorff, and let $p,q \in X$ where $p \neq q$. Since X is a Hausdorff space, there exists $G \in \mathcal{O}_p$ and $H \in \mathcal{O}_q$ such that $p \in G$, $q \in H$, and $G \cap H = \varnothing$. It follows from the definition of monad that $\mu(p) \subseteq {}^*G$ and $\mu(q) \subseteq {}^*H$. By Theorem 1–7.8 (or by the Transfer Principle), ${}^*G \cap {}^*H = \varnothing$. Hence $\mu(p) \cap \mu(q) = \varnothing$.

Conversely, suppose that X is not Hausdorff. Then there are points $p,q \in X$ such that

$$\models (\forall \mathbf{G} \in \mathcal{O}_p)(\forall \mathbf{H} \in \mathcal{O}_q)(\exists \mathbf{x} \in \mathbf{G})(\mathbf{x} \in \mathbf{H}).$$

By the previous theorem we can find internal sets $D_1 \in {}^*\mathcal{O}_p$ and $D_2 \in {}^*\mathcal{O}_q$ such that $D_1 \subseteq \mu(p)$ and $D_2 \subseteq \mu(q)$. Using the Transfer Principle, there is an x in $D_1 \cap D_2$. Hence $\mu(p) \cap \mu(q) \neq \varnothing$. ■

We next derive nonstandard equivalents of various topological notions.

THEOREM 1.3. G is open if and only if for all $p \in G$, $\mu(p) \subseteq {}^*G$.

PROOF. Let G be open, and let $p \in G$. Since $G \in \mathcal{O}_p$ and

$$\mu(p) = \bigcap \{{}^*A | A \in \mathcal{O}_p\},$$

it follows that $\mu(p) \subseteq {}^*G$.

Now let p be a point of G such that $\mu(p) \subseteq {}^*G$. Then there exists an internal set $D \in {}^*\mathcal{O}_p$ such that $D \subseteq \mu(p) \subseteq {}^*G$. Thus

$$* \models (\exists \mathbf{D} \in {}^*\mathcal{O}_p)(\mathbf{D} \subseteq {}^*\mathbf{G}).$$

Applying the Transfer Principle, we see that there is an open neighborhood A_p of p, such that $A_p \subseteq G$.

Hence if $\mu(p) \subseteq {}^*G$ for all $p \in G$, there is such an open neighborhood $A_p \subseteq G$ for each point $p \in G$. It then follows that $G = \bigcup_{p \in G} A_p$ is open. ■

Another way of writing this is

COROLLARY 1.4. G is open if and only if for all $p \in G$

$$q \approx p \quad \text{implies} \quad q \in {}^*G.$$

It is an important consequence of Theorem 1.3 that the monads determine the topology, for the monads determine the open sets and hence the topology. It follows that every topological property must be definable in terms of the monads.

We recall that F is called *closed* if $X - F$ is open. Thus F is closed if and only if for all $p \in X - F$

$$q \approx p \quad \text{implies} \quad q \in {}^*(X - F).$$

Since (by Theorem 1–7.8) ${}^*(X - F) = {}^*X - {}^*F$, this condition is equivalent to

$$p \in X \quad \text{and} \quad q \approx p \quad \text{and} \quad q \in {}^*F \quad \text{implies} \quad p \in F.$$

So we have proved

COROLLARY 1.5. F is closed if and only if for all $p \in X$

$$q \in {}^*F \quad \text{and} \quad q \in \mu(p) \quad \text{implies} \quad p \in F.$$

DEFINITION. For a topological space X, a point $p \in {}^*X$ is called *near standard* if $p \approx q$ for some $q \in X$; otherwise, p is called *remote*.

We recall that a set $K \subseteq X$ is called *compact* if every open covering of K has a finite subcovering (i.e., $K \subseteq \bigcup_{A \in \mathscr{M}} A$, $\mathscr{M} \subseteq \mathscr{O}$ implies $K \subseteq A_1 \cup \cdots \cup A_k$, for some $A_1, \ldots, A_k \in \mathscr{M}$).

The nonstandard condition for compactness is especially useful.

THEOREM 1.6. K is compact if and only if for every $p \in {}^*K$ there exists a $q \in K$ such that $p \approx q$.

An immediate consequence of this is

COROLLARY 1.7. The space X is compact if and only if every $p \in {}^*X$ is near standard.

PROOF OF THE THEOREM. Let K be compact. Suppose $p \in {}^*K$, but that $p \notin \mu(q)$ for any $q \in K$. This means that for each $q \in K$, $p \notin \bigcap\{{}^*G | G \in \mathscr{O}_q\}$. It follows that for each $q \in K$, there is some $G_q \in \mathscr{O}_q$ such that $p \notin {}^*G_q$. Now, since $q \in G_q$,

$$K \subseteq \bigcup_{q \in K} G_q.$$

Hence since K is compact, there exist $q_1, \ldots, q_l \in K$ such that we have $K \subseteq G_{q_1} \cup G_{q_2} \cup \cdots \cup Q_{q_l}$, that is,

$$\models (\forall \mathbf{x} \in \mathbf{K})(\mathbf{x} \in \mathbf{G}_{q_1} \vee \cdots \vee \mathbf{x} \in \mathbf{G}_{q_l}).$$

Using the Transfer Principle and the fact that $p \in {}^*K$, we conclude that p belongs to one of ${}^*G_{q_1}, \ldots, {}^*G_{q_l}$. But this is a contradiction because p does not belong to any ${}^*G_q$.

Conversely, suppose that for every $p \in {}^*K$, $p \approx q$ for some $q \in K$, but that K is not compact. Then there is an open covering $\mathscr{G}$ of K that cannot be reduced to a finite subcovering. Let

$$r = \{\langle X,y \rangle | X \in \mathscr{G}, y \in K, y \notin X\}.$$

We show that r is concurrent. We first note that for each $G \in \mathscr{G}$, there is a $p \in K$ such that $\langle G,p \rangle \in r$. (Otherwise, $\{G\}$ would by itself be a finite subcovering of $\mathscr{G}$.) Thus, $\mathrm{dom}(R) = \mathscr{G}$. Now, let $G_1, \ldots, G_m \in \mathscr{G}$. Since $\{G_1, \ldots, G_m\}$ cannot cover K, there is a $p \in K$ such that $p \notin G_1 \cup \cdots \cup G_m$. That is,

$$\langle G_i,p \rangle \in r, \qquad i = 1,2, \ldots, m;$$

so r is concurrent.

Applying the Concurrence Theorem, there is a $p \in {}^*U$ such that for every $G \in \mathscr{G}$, $\langle {}^*G,p \rangle \in {}^*r$. In fact, since ${}^*r \subseteq {}^*\mathscr{G} \times {}^*K$, $p \in {}^*K$. Now, for each $G \in \mathscr{G}$,

$$\models (\forall \mathbf{x} \in \mathbf{K})(\langle \mathbf{G},\mathbf{x} \rangle \in \mathbf{r} \rightarrow \neg(\mathbf{x} \in \mathbf{G})).$$

By the Transfer Principle, we have that for every $G \in \mathscr{G}$, $p \notin {}^*G$. By hypothesis, there exists a $q \in K$ such that $p \approx q$. Since $\mathscr{G}$ is an open covering of K and since $q \in K$, for *some* $G \in \mathcal{G}$, $q \in G$. Hence we have that $p \in \mu(q) \subseteq {}^*G$, since $\mu(q)$ is the intersection of all ${}^*G$ with $G \in \mathcal{O}_q$. Hence we have that both $p \in {}^*G$ and $p \notin {}^*G$: a contradiction. ■

We conclude this section with nonstandard proofs of a pair of elementary theorems.

THEOREM 1.8. A closed subset of a compact set is compact.

PROOF. Let $F \subseteq K$, where F is closed and K is compact. Let $p \in {}^*F$. Then $p \in {}^*K$. Since K is compact, $p \approx q$ for some point $q \in K$. But since F is closed, it follows that $q \in F$. By Theorem 1.6, F is compact. ■

THEOREM 1.9. In a Hausdorff space, a compact set is closed.

PROOF. Let K be compact. Let $q \in {}^*K$ and let $q \in \mu(p)$. We wish to show that $p \in K$. Since K is compact and $q \in {}^*K$, it follows that $q \in \mu(r)$ for some $r \in K$. Since the space is Hausdorff, we conclude that $r = p$. ■

2. MAPPINGS AND PRODUCTS

We now consider situations involving more than one topological space. In such a case, the notation $\mu(p)$ is ambiguous; however, the context will in all cases resolve the ambiguity.

Let f map the topological space X into the topological space Y. Then f is *continuous at* $p \in X$ means that if $G \subseteq Y$ is any open neighborhood of $f(p)$, then there is an open neighborhood H of p such that $f[H] \subseteq G$.

Of course, in order to discuss continuity from a nonstandard point of view, we must assume that $X, Y \in U$. (Unless $X, Y \subseteq S$, we cannot assume that ${}^*f$ is an extension of f. Of course, ${}^*f$ maps ${}^*X$ into ${}^*Y$.)

THEOREM 2.1. Let f map X into Y where X and Y are both topological spaces. Let $p \in X$. Then f is continuous at p if and only if

$$q \approx p \quad \text{implies} \quad {}^*f(q) \approx f(p).$$

This last condition may be equivalently written:

$$q \in \mu(p) \quad \text{implies} \quad {}^*f(q) \in \mu(f(p)),$$

or simply,

$${}^*f[\mu(p)] \subseteq \mu(f(p)).$$

PROOF. Let f be continuous at p. Let G be any open neighborhood of $f(p)$. Then there exists an open neighborhood $H \in \mathcal{O}_p$ such that $f[H] \subseteq G$. By Theorems 1–7.11 and 1–7.13, ${}^*f[{}^*H] \subseteq {}^*G$. Since $\mu(p) \subseteq {}^*H$, it follows that

$${}^*f[\mu(p)] \subseteq {}^*f[{}^*H] \subseteq {}^*G.$$

Hence

$$^*f[\mu(p)] \subseteq \bigcap\{^*G | G \in \mathcal{O}_{f(p)}\} = \mu(f(p)).$$

Proceeding in the converse direction, by Theorem 1.1 there exists an internal set D such that $D \in {}^*\mathcal{O}_p$ and $D \subseteq \mu(p)$. Hence by hypothesis, $q \in D$ implies $^*f(q) \in \mu(f(p))$. So, if H is any open neighborhood of $f(p)$, then $q \in D$ implies $^*f(q) \in {}^*H$. We conclude that for each $H \in \mathcal{O}_{f(p)}$,

$$* \models (\exists \mathbf{D} \in {}^*\mathcal{O}_p)(\forall \mathbf{x} \in \mathbf{D})(^*\mathbf{f}(\mathbf{x}) \in {}^*\mathbf{H}).$$

By the Transfer Principle, there is an open neighborhood D of p such that $f[D] \subseteq H$. Thus f is continuous at p. ■

As a simple application of this theorem, we give a nonstandard proof of a basic theorem.

COROLLARY 2.2. If f is continuous at each point of K and K is compact, then $f[K]$ is compact.

PROOF. Let $q \in {}^*(f[K])$. By Theorem 1–7.13, $^*(f[K]) = {}^*f[^*K]$, so that $q = {}^*f(r)$ for some $r \in {}^*K$. Since K is compact, $r \approx r_0$ for some $r_0 \in K$. By Theorem 2.1, $q = {}^*f(r) \approx f(r_0) \in f[K]$. By Theorem 1.6, $f[K]$ is compact. ■

We wish to deal next with Cartesian products of topological spaces, and for this purpose we recall that if $\mathcal{O}$ is the collection of open sets of the topological space X, then $\mathcal{G}$ is called a *basis* of X if

(1) $\mathcal{G} \subseteq \mathcal{O}$;
(2) for each $G \in \mathcal{O}$ there is a family $\{G_i | i \in J\}$ such that each $G_i \in \mathcal{G}$ and $G = \bigcup_{i \in J} G_i$.

If $\mathcal{G}$ is a basis for X and $p \in X$, we write

$$\mathcal{G}_p = \mathcal{G} \cap \mathcal{O}_p,$$

that is,

$$\mathcal{G}_p = \{G \in \mathcal{G} | p \in G\}.$$

THEOREM 2.3. If $\mathcal{G}$ is a basis for X, then for each $p \in X$,

$$\mu(p) = \bigcap_{G \in \mathcal{G}_p} {}^*G.$$

PROOF.

$$\mu(p) = \bigcap_{G \in \mathcal{O}_p} {}^*G \subseteq \bigcap_{G \in \mathcal{G}_p} {}^*G.$$

Conversely, for each $H \in \mathcal{O}_p$, we can find a $G_H \in \mathcal{G}_p$ such that $G_H \subseteq H$. (This can be done because H is the union of sets belonging to $\mathcal{G}$, and at least one of these sets must contain p.) Thus

$$\bigcap_{G \in \mathcal{G}_p} {}^*G \subseteq \bigcap_{H \in \mathcal{O}_p} {}^*G_H \subseteq \bigcap_{H \in \mathcal{O}_p} {}^*H = \mu(p). \quad \blacksquare$$

Now we consider the product

$$X = \prod_{i \in J} X_i = \{f | f(i) \in X_i \quad \text{for each} \quad i \in J\},$$

where we assume that $\{X_i | i \in J\} \in U$. If each X_i is a topological space, then the *product topology* on X, as usually defined, is characterized by the existence of a *basis* that consists precisely of all sets of the form,

$$A = \{f | f(v_i) \in G_i, \qquad i = 1,2, \ldots, l\},$$

where $G_1, \ldots, G_l$ are open sets in $X_{v_1}, \ldots, X_{v_l}$, respectively. From a nonstandard point of view, we have the more intuitive characterization of the product topology:

THEOREM 2.4. Let $X = \prod_{i \in J} X_i$. Let $g \in {}^*X$ and $f \in X$. Then $g \approx f$ if and only if $g(i) \approx f(i)$ for all $i \in J$.

PROOF. In this proof, we write $\mathcal{G}$ for the basis of X described above.

Let $g \approx f$, but suppose that for some $j_0 \in J$, $g(j_0) \not\approx f(j_0)$, that is, $g(j_0) \notin \mu(f(j_0))$. By the definition of monad, $g(j_0) \notin {}^*G$ for some $G \in \mathcal{O}_{f(j_0)}$. Let $W = \{h \in X | h(j_0) \in G\}$, so that W is an open set. (In fact, $W \in \mathcal{G}$.) Now,

$$\models (\forall \mathbf{t} \in \mathbf{X})(\mathbf{t} \in \mathbf{W} \rightarrow (\mathbf{t} \upharpoonright \mathbf{j}_0) \in \mathbf{G}).$$

By the Transfer Principle,

$$* \models (\forall \mathbf{t} \in {}^*\mathbf{X})(\mathbf{t} \in {}^*\mathbf{W} \rightarrow (\mathbf{t} \upharpoonright \mathbf{j}_0) \in {}^*\mathbf{G}).$$

Since $g \in {}^*X$ and $g(j_0) = g \upharpoonright j_0 \notin {}^*G$, we must have $g \notin {}^*W$. On the other hand, $f \in W$ so that $W \in \mathcal{O}_f$. Hence by definition of $\mu(f)$, $g \notin \mu(f)$, that is, $g \not\approx f$.

Conversely, let $g(i) \approx f(i)$ for all $i \in J$. Let $G \in \mathcal{G}_f$. Then we may write $G = \{h | h(v_i) \in G_i, i = 1,2, \ldots, l\}$ where each G_i is an open set in X_{v_i}. Since $f \in G$, it follows that $f(v_i) \in G_i$, $i = 1, \ldots, l$. By hypothesis, $g(v_i) \approx f(v_i)$. Since each G_i is an open set, $g(v_i) \in {}^*G_i$. It follows using the Transfer Principle that $g \in {}^*G$. Hence using Theorem 2.3, $g \in \bigcap_{G \in \mathcal{G}_f} {}^*G = \mu(f)$, that is, $g \approx f$. $\blacksquare$

THEOREM 2.5. The product of Hausdorff spaces is Hausdorff.

PROOF. Let $X = \prod_{i \in J} X_i$ where each X_i is Hausdorff. Let $f,g \in X$ where $f \neq g$. We wish to show that $\mu(f) \cap \mu(g) = \varnothing$. Suppose, instead, that $h \in \mu(f) \cap \mu(g)$. That is, $h(i) \approx f(i)$, $h(i) \approx g(i)$ for all i in J. But since each

X_i is a Hausdorff space, it follows that for all $i \in J$, $f(i) = g(i)$, that is, $f = g$. ■

THEOREM 2.6 (TYCHONOFF'S THEOREM). If X_i is compact for each $i \in J$, then so is $\prod_{i \in J} X_i$.

PROOF. Let $X = \prod_{i \in J} X_i$, and let $g \in {}^*X$. We wish to show that $g \approx f$ for some $f \in X$. Since $g \in {}^*X$, we have by the Transfer Principle that $g(i) \in {}^*X_i$ for all $i \in J$. By the compactness of each X_i, $g(i) \approx a_i$ for some $a_i \in X_i$. Thus the mapping f defined by $f(i) = a_i$ for $i \in J$ is in X. Clearly, $g(i) \approx f(i)$ for all $i \in J$, and by Theorem 2.4, $g \approx f$. ■

A technical point: if the X_is are all Hausdorff, then each a_i is uniquely determined, and this proof does not explicitly use the axiom of choice; otherwise, the axiom of choice is used in obtaining f. Of course, a weak form of the axiom of choice has already been used in our development.

3. TOPOLOGICAL GROUPS

A *topological group* is a set G that simultaneously satisfies:

(1) G is a *topological space*; that is, a collection $\mathcal{O}$ of subsets of G, satisfying the usual requirements, has been singled out as the *open sets* of G.
(2) G is a *group* under the operation $\cdot$, with identity e, and with the inverse of $x \in G$ being x^{-1} (As usual we write xy for $x \cdot y$).
(3) The functions $x \cdot y$ and x^{-1} are *continuous* in the topology defined by $\mathcal{O}$.

We assume that $G \subseteq S$, so that $\mathcal{O}$ and the product and inverse functions each belongs to the standard universe U.

Applying the previous discussion, the continuity requirements of (3) amount to

(3a) If $x_0, y_0 \in G$, $x, y \in {}^*G$, $x \approx x_0$, $y \approx y_0$, then $x \cdot y \approx x_0 \cdot y_0$.
(3b) If $x_0 \in G$, $x \in {}^*G$, $x \approx x_0$, then $x^{-1} \approx x_0^{-1}$.

Since G is a group, we have

$$(\forall \mathbf{x} \in \mathbf{G})(\forall \mathbf{y} \in \mathbf{G})(\forall \mathbf{z} \in \mathbf{G})((\mathbf{x} \cdot (\mathbf{y} \cdot \mathbf{z})) = ((\mathbf{x} \cdot \mathbf{y}) \cdot \mathbf{z}) \,\&\, ((\mathbf{e} \cdot \mathbf{x}) = \mathbf{x}) \,\&\, (\mathbf{x}^{-1} \cdot \mathbf{x} = \mathbf{e})).$$

It then follows from the Transfer Principle that ${}^*G$ is also a group. Since we are assuming that $G \subseteq S$, we have $G \subseteq {}^*G$ so that G is a subgroup of ${}^*G$.

Let J be the set of near-standard elements of ${}^*G$; that is,

$$J = \{x \in {}^*G | x \approx x_0 \text{ for some } x_0 \in G\}.$$

THEOREM 3.1. J is a subgroup of $^*G$.

PROOF. Let $x,y \in J$. It suffices to show that $x \cdot y \in J$ and $x^{-1} \in J$. Let $x \approx x_0$ and $y \approx y_0$ where $x_0,y_0 \in G$. By (3a), (3b) above,

$$x \cdot y \approx x_0 \cdot y_0 \in G, \qquad x^{-1} \approx x_0^{-1} \in G.$$

Hence $x \cdot y$, x^{-1} belong to J. ■

THEOREM 3.2. $\mu(e)$ is a normal subgroup of J.

PROOF. Clearly, $\mu(e) \subseteq J$, since $e \in G$. Let $x,y \in \mu(e)$. Then $x \approx e$ and $y \approx e$, and so $x \cdot y \approx e \cdot e = e$ and $x^{-1} \approx e^{-1} = e$. Hence $\mu(e)$ is a subgroup of J.

Next let $y \in J$, $x \in \mu(e)$. We wish to show that $y^{-1}xy \in \mu(e)$. We have $y \approx y_0$ for some $y_0 \in G$ and $x \approx e$. Hence

$$y^{-1}xy \approx y_0^{-1}ey_0 = e. \quad ■$$

Let $A,B \subseteq {}^*G$ and $x \in {}^*G$. As usual, we write

$$xA = \{xy | y \in A\},$$
$$AB = \{xy | x \in A \quad \text{and} \quad y \in B\}.$$

THEOREM 3.3. For $x \in G$,

$$\mu(x) = x\mu(e) = \mu(e)x.$$

PROOF. Let $t \in x\mu(e)$, that is, $t = xy$ for some $y \approx e$. Then $t \approx xe = x$, that is, $t \in \mu(x)$. Conversely, if $t \in \mu(x)$, then $t \approx x$, so that $x^{-1}t \approx x^{-1}x = e$. Then $x^{-1}t \in \mu(e)$, so that $t = x(x^{-1}t) \in x\mu(e)$. This proves that $\mu(x) = x\mu(e)$.

That $\mu(x) = \mu(e)x$ can be proved similarly; or we can note that since $\mu(e)$ is normal in J, the left coset $x\mu(e)$ is the same as the right coset $\mu(e)x$. ■

LEMMA 1. Let $y \approx x \in G$. Then $y\mu(e) = x\mu(e) = \mu(x)$.

PROOF. $y = ye$ so $y \in y\mu(e)$. Also, $y \in \mu(x) = x\mu(e)$. But two cosets with an element in common must be identical. ■

Thus we have seen that the cosets of $\mu(e)$ in J are precisely the monads $\mu(x)$ for $x \in G$.

LEMMA 2. $\mu(xy) = \mu(x)\mu(y)$ for $x,y \in G$.

PROOF. If $z \in \mu(xy) = xy\mu(e)$, then $z = xyt$ where $t \approx e$. But $x \in \mu(x)$ and $yt \in y\mu(e) = \mu(y)$. So $z \in \mu(x)\mu(y)$.

Conversely, if $z \in \mu(x)\mu(y)$, then $z = uv$ where $u \approx x$, $v \approx y$. Then $z \approx xy$, that is, $z \in \mu(xy)$. ■

Now the elements of quotient group $J/\mu(e)$ are precisely the cosets $\mu(x)$. Hence Lemma 2 yields

THEOREM 3.4. The mapping $x \to \mu(x)$ of G onto $J/\mu(e)$ is a homomorphism.

If G is a Hausdorff space, then each coset $\mu(x)$ contains just one element of G. That is, the map $x \to \mu(x)$ is one-one in this case.

This proves

THEOREM 3.5. If G is Hausdorff, then G is isomorphic to $J/\mu(e)$.

THEOREM 3.6. Let $H \subseteq G$, and let $x \in G$. Then if H is an open set, so is xH.

PROOF. Let $xh \in xH$ where $h \in H$. Since H is open, $\mu(h) \subseteq {}^*H$. To show that xH is open, it suffices to prove that $\mu(xh) \subseteq {}^*(xH) = x({}^*H)$. So, let $q \approx xh$. Then $x^{-1}q \approx h$. Hence $x^{-1}q \in {}^*H$ and $q \in x({}^*H)$. ∎

THEOREM 3.7. Let $A,B \subseteq G$ where A is closed and B is compact. Then $A \cdot B$ is closed.

PROOF. Let $t \in {}^*(A \cdot B)$ where $t \approx g$ for some $g \in G$. It suffices to show that $g \in A \cdot B$. Since ${}^*(A \cdot B) = {}^*A \cdot {}^*B$, we can write $t = \alpha \cdot \beta$, for $\alpha \in {}^*A$, $\beta \in {}^*B$. Since B is compact, $\beta \approx b$, for some $b \in B$. Hence $t = \alpha\beta \approx \alpha b$. Since $t \approx g$, it follows that $\alpha b \approx g$, and $\alpha \approx gb^{-1}$. Since $gb^{-1} \in G$, $\alpha \in {}^*A$, and A is closed, we have $gb^{-1} \in A$. Thus $g = (gb^{-1})b \in A \cdot B$. ∎

4. THE EXISTENCE OF HAAR MEASURE[1]

We next show how nonstandard methods can be used to give a simple and intuitive treatment of the key step in the construction of Haar measure on a locally compact Hausdorff topological group. (We assume no previous knowledge of these matters.)

In this section, G is a Hausdorff topological group; as usual, $\mathcal{O}$ is the collection of open sets of G, and for each $q \in G$,

$$\mathcal{O}_q = \{E \in \mathcal{O} | q \in E\}.$$

DEFINITION. $A \subseteq G$ is *bounded* if $A \subseteq K$ for some compact set K.

Let $\mathcal{B}$ be the collection of all bounded subsets of G.

DEFINITION. G is called *locally compact* if $\mathcal{B} \cap \mathcal{O}_e \neq \emptyset$.

We now assume that in fact G *is locally compact*. Hence there exists some A_0 such that $A_0 \in \mathcal{B}$ and $A_0 \in \mathcal{O}_e$. We fix some such set A_0 in what follows.

[1] We follow Hausner [6]; compare also with Parikh [13].

DEFINITION. *H separates A from B* if

(1) $H \in \mathcal{O}_e$ and $A, B \in \mathcal{B}$;
(2) For all $x \in G$ either $A \cap xH = \varnothing$ or $B \cap xH = \varnothing$.

H separates A from B means that no matter how H is left "translated," the result never simultaneously intersects both A and B.

DEFINITION. A and B are *separated* if some H separates them.

We are going to be concerned with functions f mapping $\mathcal{B}$ into some ordered field D.

DEFINITION. Let f map $\mathcal{B}$ into the ordered field D. Then f is called *admissible* if the following are satisfied:

(a) $f(A) \geq 0$ for all $A \in \mathcal{B}$.
(b) $A \subseteq B$ implies $f(A) \leq f(B)$.
(c) $f(A \cup B) \leq f(A) + f(B)$.
(d) $f(A_0) = 1$.
(e) $f(xA) = f(A)$ for all $x \in G$ and $A \in \mathcal{B}$.

DEFINITION. f is *weakly additive* if $f(A \cup B) = f(A) + f(B)$ whenever A and B are separated.

Finally,

DEFINITION. A function λ that maps $\mathcal{B}$ into R (the set of real numbers) is a *content function* (or simply a *content*) on G if λ is both admissible and weakly additive.

Our main theorem is

THEOREM 4.1. There is a content function on G.

Actually, a Haar measure is a map ν of the "Borel" sets of G into $R \cup \{\infty\}$ such that

(1) $\nu(A) \geq 0$ whenever defined.
(2) $B \in \mathcal{B}$ implies $\nu(B) < \infty$.
(3) $\nu(A_0) = 1$.
(4) $\nu(xA) = \nu(A)$ for $x \in G$.
(5) ν is "countably additive."

Given a *content* on G, the construction of such a v is a straightforward application of the methods of measure theory.[1]

Now, let $A \in \mathscr{B}$, $H \in \mathcal{O}_e$. Then $A \subseteq K$ where K is compact. Since $e \in H$, $x \in xH$. Also, by Theorem 3.6 xH is open for each $x \in G$. Hence $\bigcup_{x \in G} (xH) = G$, so that $\{xH | x \in G\}$ is an open covering of K. Since K is compact and $A \subseteq K$,

$$A \subseteq x_1 H \cup x_2 H \cup \cdots \cup x_N H \qquad (*)$$

for some finite set $\{x_1, \ldots, x_N\}$. Let $(A:H)$ be the least value of N for which (*) holds, and we define

$$g(A,H) = g_H(A) = \frac{(A:H)}{(A_0:H)}$$

for each $A \in \mathscr{B}$.

LEMMA 1. For each $H \in \mathcal{O}_e$, g_H is admissible. Moreover, if H separates A from B, then

$$g_H(A \cup B) = g_H(A) + g_H(B).$$

PROOF. We verify (a) to (e) from p. 85.

(a) The proof is obvious.
(b) If $A \subseteq B$, then $B \subseteq x_1 H \cup \cdots \cup x_N H$ implies $A \subseteq x_1 H \cup \cdots \cup x_N H$, so $(A:H) \leq (B:H)$.
(c) If $A \subseteq x_1 H \cup \cdots \cup x_N H$ and $B \subseteq y_1 H \cup \cdots \cup y_K H$, then $A \cup B \subseteq x_1 H \cup \cdots \cup x_N H \cup y_1 H \cup \cdots \cup y_K H$, so $(A \cup B:H) \leq (A:H) + (B:H)$.
(d) $g_H(A_0) = (A_0:H)/(A_0:H) = 1$.
(e) If $A \subseteq x_1 H \cup \cdots \cup x_N H$, then $xA \subseteq (xx_1)H \cup \cdots \cup (xx_N)H$, whereas if $xA \subseteq y_1 H \cup \cdots \cup y_K H$, then $A \subseteq (x^{-1}y_1)H \cup \cdots \cup (x^{-1}y_K)H$. Hence $(xA:H) = (A:H)$.

Finally, let H separate A from B, and let

$$A \cup B \subseteq x_1 H \cup \cdots \cup x_N H.$$

Since no $x_i H$ can meet both A and B, we can write (reordering the x_is if necessary)

$$A \subseteq x_1 H \cup \cdots \cup x_K H; \qquad B \subseteq x_{K+1} H \cup \cdots \cup x_N H.$$

Hence $(A \cup B:H) = (A:H) + (B:H)$. ■

LEMMA 2. There is an admissible weakly additive function k that maps $\mathscr{B}$ into $^*R$.

PROOF. The mapping g above maps $\mathscr{B} \times \mathcal{O}_e$ into R. Hence $^*g$ maps $^*\mathscr{B} \times {}^*\mathcal{O}_e$ into $^*R$. Now, we know by Theorem 1.1 that there exists an

[1] The content is first used to define an "outer measure." For details, see, for example, Halmos [4].

internal set $D \in {}^*\mathcal{O}_e$ such that $D \subseteq \mu(e)$. We define $k(A) = {}^*g_D(A)$ for all $A \in \mathscr{B}$. Since we know that for each $H \in \mathcal{O}_e$, g_H is admissible, it follows from the Transfer Principle that $k = {}^*g_D$ is admissible. We now show that k is weakly additive. Let A and B be separated by $H \in \mathcal{O}_E$. We have to show that $k(A \cup B) = k(A) + k(B)$. Since any open neighborhood of e which is a subset of H also separates A from B, we have

$$\models (\forall \mathbf{L} \in \mathcal{O}_e)(\mathbf{L} \subseteq \mathbf{H} \rightarrow \mathbf{g}(\mathbf{A} \cup \mathbf{B},\mathbf{L}) = \mathbf{g}(\mathbf{A},\mathbf{L}) + \mathbf{g}(\mathbf{B},\mathbf{L})).$$

Applying the Transfer Principle,

$$*\models (\forall \mathbf{L} \in {}^*\mathcal{O}_e)(\mathbf{L} \subseteq {}^*\mathbf{H} \rightarrow {}^*\mathbf{g}(\mathbf{A} \cup \mathbf{B},\mathbf{L}) = {}^*\mathbf{g}(\mathbf{A},\mathbf{L}) + {}^*\mathbf{g}(\mathbf{B},\mathbf{L})).$$

But since $D \subseteq \mu(e) \subseteq {}^*H$, and $D \in {}^*\mathcal{O}_e$, we have

$${}^*g_D(A \cup B) = {}^*g_D(A) + {}^*g_D(B),$$

that is,

$$k(A \cup B) = k(A) \cup k(B). \quad \blacksquare$$

Finally, we have

LEMMA 3. $k(A)$ is finite for each $A \in \mathscr{B}$.

PROOF. Let A be any member of $\mathscr{B}$. Then A is bounded, and thus $A \subseteq K$ for K compact. Since K is compact and $A \subseteq K \subseteq \bigcup_{x \in G} xA_0$,

$$A \subseteq x_1A_0 \cup x_2A_0 \cup \cdots \cup x_nA_0$$

for some finite set $\{x_1,x_2, \ldots ,x_n\}$. Using the fact that k is admissible,

$$\begin{aligned} k(A) &\leq k(x_1A_0 \cup \cdots \cup x_nA_0) \\ &\leq k(x_nA_0) + \cdots + k(x_nA_0) \\ &= k(A_0) + \cdots + k(A_0) = n. \quad \blacksquare \end{aligned}$$

We now define the desired content function by

$$\lambda(A) = {}^\circ(k(A)),$$

so that λ maps $\mathscr{B}$ into R. Since $^\circ$ is a homomorphism that preserves $\leq$, λ is admissible and weakly additive, so that it is indeed a content.

5. METRIC SPACES

We recall that a *metric* on a set $X \neq \varnothing$ is a mapping $\rho: X \times X \rightarrow R$ such that for all $x,y,z \in X$,

(a) $\rho(x,y) \geq 0$,
(b) $\rho(x,y) = 0$ if and only if $x = y$,
(c) $\rho(x,y) = \rho(y,x)$,
(d) $\rho(x,z) \leq \rho(x,y) + \rho(y,z)$.

$\rho(x,y)$ is called the *distance* between x and y. The set X together with the metric ρ is called a *metric space.*

DEFINTIION. For $p \in X$ and real $r > 0$, the set

$$B_p(r) = \{q \in X | \rho(q,p) < r\}$$

is called the *open ball* with *center* p and radius r. If $G \subseteq X$, and if for each $p \in G$, there is $r \in R^+$, such that $B_p(r) \subseteq G$, then G is called an *open set* in the metric space X.

Writing $\mathcal{O}$ for the collection of all such open sets, it is an entirely trivial matter to see that X with the open sets $\mathcal{O}$ is a topological space. Hence all previous considerations regarding topological spaces apply. We have

LEMMA 1. For each $p \in X$ and each $r \in R^+$, $B_p(r)$ is an open set.

PROOF. Let $q \in B_p(r)$, that is, $\rho(q,p) < r$. Let $s = r - \rho(q,p) > 0$. We claim that $B_q(s) \subseteq B_p(r)$, which, of course, gives the result. To verify the claim, let $m \in B_q(s)$, so that $\rho(m,q) < s$. Then

$$\rho(m,p) \leq \rho(m,q) + \rho(q,p) < s + \rho(q,p) = r.$$

Hence $m \in B_p(r)$. ■

We assume that $X,R \subseteq S$, so that $\rho \in U$ and $^*\rho$ maps $^*X \times {}^*X$ into $^*R$, and we write ρ instead of $^*\rho$ as usual. By the Transfer Principle, (a) to (d) hold for $x,y,z \in {}^*X$. Of course, the "distance" between two points of $^*X$ is in general a hyperreal number.

THEOREM 5.1. For any $p \in X$,

$$\mu(p) = \bigcap_{r \in R^+} {}^*B_p(r).$$

PROOF. By Lemma 1, $B_p(r) \subseteq \mathcal{O}_p$ for all $r \in R^+$. Hence

$$\mu(p) = \bigcap_{G \in \mathcal{O}_p} {}^*G \subseteq \bigcap_{r \in R^+} {}^*B_p(r).$$

For the opposite inclusion, consider any $G \in \mathcal{O}_p$. By definition, there is an $r = r_G \in R^+$ such that $B_p(r_G) \subseteq G$. Then

$$\bigcap_{r \in R^+} {}^*B_p(r) \subseteq \bigcap_{G \in \mathcal{O}_p} {}^*B_p(r_G) \subseteq \bigcap_{G \in \mathcal{O}_p} {}^*G = \mu(p). \quad ■$$

For $X = R$, the usual metric is $\rho(x,y) = |x - y|$. With this metric, and for some fixed $x_0 \in R$,

$$B_r(x_0) = \{x \in R | x_0 - r < x < x_0 + r\},$$

that is, $B_r(x_0) = I_r$ in the sense of the very beginning of this chapter. Hence Theorem 5.1 shows that $\mu(x_0) = \bigcap_{r \in R^+} {}^*I_r$, as was anticipated.

Returning to an arbitrary metric space, let $q \in {}^*X$, $p \in X$. Then by Theorem 5.1, $q \approx p$ if and only if $0 \leq \rho(q,p) \leq r$ for all $r \in R^+$, that is, if and only if $\rho(q,p)$ is infinitesimal. This suggests that we write $q \approx p$ for arbitrary $q,p \in {}^*X$ to mean $\rho(q,p)$ is infinitesimal. In particular, for $X = R$ and $x,y \in R$, we have

$$x \approx y \text{ if and only if } |x - y| \text{ is infinitesimal,}$$

so that the notation agrees with that of Chapter 2.

For any $p \in {}^*X$ we now write $\mu(p) = \{q \in {}^*X | \rho(q,p) \approx 0\}$, so that we can speak of the monad of any point of ${}^*X$.

DEFINITION. $p \in {}^*X$ is called *finite* if for some $q \in X$, $\rho(p,q)$ is finite.

COROLLARY 5.2 Every near-standard point is finite.

PROOF. Let $p \in {}^*X$ be near-standard. Then, for some $q \in X$, $p \approx q$, that is, $\rho(p,q) \approx 0$. Thus $\rho(p,q)$ is finite. ■

COROLLARY 5.3. A metric space is Hausdorff.

PROOF. Let $p,q \in X$, and let $s \in \mu(p)$, $s \in \mu(q)$. So $\rho(s,p)$, $\rho(s,q) \approx 0$. Hence $\rho(p,q) \leq \rho(p,s) + \rho(s,q) \approx 0$. Since $\rho(p,q) \in R$, $\rho(p,q) = 0$ and $p = q$. ■

This justifies:

DEFINITION. For any near-standard point p, ${}^\circ p$ is the unique $q \in X$ such that $p \in \mu(q)$. ${}^\circ p$ is called the *standard part* of p.

Of course, this agrees with the notation ${}^\circ x$ for x a finite hyperreal number.

DEFINITION. $B \subseteq X$ is *bounded* if $B \subseteq B_p(r)$ for some $p \in X$ and $r \in R^+$.

LEMMA 2. If $p \in {}^*X$ is finite and $q \in X$, then $\rho(p,q)$ is finite.

PROOF. Since $p \in {}^*X$ is finite, $\rho(p,s)$ is finite for some $s \in X$. Let q be any other standard point. Since

$$\rho(p,q) \leq \rho(p,s) + \rho(s,q)$$

and $\rho(p,s)$ and $\rho(s,q)$ are finite, it follows that $\rho(p,q)$ is finite. ■

THEOREM 5.4. $B \subseteq X$ is bounded if and only if every element of ${}^*B$ is finite.

PROOF. Let $B \subseteq B_p(r)$ for $p \in X$ and $r \in R^+$. Then

$$\models (\forall \mathbf{x} \in \mathbf{B})(\rho(\mathbf{x},\mathbf{p}) < \mathbf{r}).$$

By the Transfer Principle, it follows that for any $q \in {}^*B$, $\rho(q,p) < r$. Hence all the elements of ${}^*B$ are finite.

Conversely, let each $x \in {}^*B$ be finite, and fix some $p \in X$; thus by Lemma 2 $\rho(x,p) < r$ for any positive infinite hyperreal r. Hence

$$* \models (\exists \mathbf{r} \in {}^*\mathbf{R}^+)(\forall \mathbf{x} \in {}^*\mathbf{B})(\rho(\mathbf{x},\mathbf{p}) < \mathbf{r}).$$

Applying the Transfer Principle, it follows that there is an $r \in R^+$ such that $B \subseteq B_p(r)$. ■

COROLLARY 5.5 A compact set is bounded.

PROOF. Let K be compact and $q \in {}^*K$. Then $q \approx p$ for some $p \in K$. But this implies that q is finite. By Theorem 5.4, K is bounded. ■

THEOREM 5.6. The space X satisfies the condition,

every bounded closed set is compact,

if and only if every finite point of ${}^*X$ is near standard.

PROOF. Let every finite point of ${}^*X$ be near standard. Let B be closed and bounded, and let $x \in {}^*B$. To see that B is compact, we show that $x \approx q$ for some $q \in B$. Since B is bounded, x must be finite; therefore, by hypothesis, x is near standard. That is, $x \approx q$ for some $q \in X$. Since B is closed, $q \in B$.

Conversely, suppose that the condition is satisfied. Let $p \in {}^*X$ be finite; we show that p is near standard. $\rho(p,q) < r$ for some $q \in X$ and $r \in R^+$. Let

$$H = \{x \in X | \rho(x,q) \leq r\}.$$

Clearly, $H \subseteq B_q(r + 1)$, and thus H is bounded. To show that H is also closed, let $t \in {}^*H$, $t \approx s \in X$. By the Transfer Principle, $\rho(t,q) \leq r$. Hence $\rho(s,q) \leq r + \rho(t,s)$. Taking standard parts, $\rho(s,q) \leq r$, so that $s \in H$; thus H is closed. Since H is both bounded and closed, by the condition it is also compact. Hence every member of ${}^*H$ is near standard. Since $p \in {}^*H$, it is near standard, which completes the proof. ■

Naturally, any subset A of a metrix space S with metric ρ is itself a metric space with metric ρ (restricted of course to A).

THEOREM 5.7. Let $K \subseteq X$, X a metric space. Then K is a compact set as a subset of X if and only if K is a compact space as a subspace of X.

PROOF. Both conditions are clearly equivalent to

For every $p \in {}^*K$, we have $p \approx q$ for some $q \in K$. ■

The significance of this theorem is that theorems about compact metric spaces can also be regarded as theorems about compact sets in a metric space.

We next consider sequences of points in metric spaces, that is, mappings s from N^+ into X, where as usual we write s_n for $s(n)$. Then $^*s$ (which we, of course, write as s) maps $^*N^+$ into $^*X$. We write $s_n \to x$ and say that x is the limit of $\{s_n | n \in N^+\}$ if $\rho(s_n,x) \to 0$. By our discussion of convergence of sequences of real numbers (Theorem 2–5.1), we immediately have

THEOREM 5.8. $s_n \to x$ if and only if $\rho(s_n,x) \approx 0$ for all $n \in {}^*N - N$.

COROLLARY 5.9. $s_n \to x$ implies $\{s_n | n \in N^+\}$ is bounded.

PROOF. Clearly, for $n \in N^+$, s_n is finite. By Theorem 5.8, s_n is near standard and hence finite for $n \in {}^*N - N$. Thus s_n is finite for all $n \in {}^*N$.

The set of terms of the sequence $\{s_n | n \in N^+\}$ is simply the range $s[N^+]$. By Theorem 1–7.13, $^*(s[N^+]) = s[^*N^+]$. Hence by Theorem 5.4, $s[N^+]$ is a bounded set. ■

COROLLARY 5.10. $s_n \to x$ and $s_n \to y$ implies $x = y$.

PROOF. For $n \in {}^*N - N$, $s_n \approx x$ and $s_n \approx y$, so that $x \approx y$. Since X is Hausdorff, $x = y$. ■

DEFINITION. $x \in X$ is a *limit point* of $\{s_n | n \in N^+\}$ if some subsequence of $\{s_n\}$ has x as a limit.

If x is a limit point of $\{s_n\}$, then, of course, for each $\varepsilon \in R^+$ the inequality

$$\rho(s_n,x) < \varepsilon$$

is satisfied for infinitely many $n \in N^+$. Conversely, if for each $\varepsilon > 0$, this inequality is satisfied for infinitely many n, then a subsequence of $\{s_n\}$ with x as its limit may be obtained as follows:

Let $n_1 = 1$, and supposing $n_1 < n_2 < \cdots < n_j$ defined, let n_{j+1} be $> n_j$ such that

$$\rho(s_{n_{j+1}},x) < \frac{1}{j}.$$

Then $s_{n_j} \to x$.

Generalizing Theorem 2–5.3, we have

THEOREM 5.11. x is a limit point of $\{s_n\}$ if and only if $s_\nu \approx x$ for some $\nu \in {}^*N - N$.

PROOF. In the proof of Theorem 2–5.3, simply replace "$|s_n - x|$" by "$\rho(s_n,x)$" and "$\mathbf{|s_n - x|}$" by "$\boldsymbol{\rho(s_n,x)}$" at all occurrences. ■

As an immediate corollary, we obtain

THEOREM 5.12 (BOLZANO–WEIERSTRASS THEOREM). In a compact metric space, every sequence has a convergent subsequence.

PROOF. Let $\{s_n | n \in N^+\}$ be a sequence of points of the space, and let $\nu \in {}^*N - N$. Since the space is compact, s_ν is near standard, that is, $s_\nu \approx x \in X$. Hence x is a limit point of the sequence, that is, some subsequence converges to x. ■

DEFINITION. The sequence $\{s_n | n \in N^+\}$ is a *Cauchy sequence* if for all real $\varepsilon > 0$ there is an $n_0 \in N^+$ such that $m,n > n_0$ implies $\rho(s_m,s_n) < \varepsilon$.

THEOREM 5.13. $\{s_n | n \in N^+\}$ is a Cauchy sequence if and only if $s_n \approx s_m$ for all $m,n \in {}^*N - N$.

PROOF. Let $\{s_n\}$ be a Cauchy sequence, and let $\varepsilon \in R^+$. Then, there is some $n_0 \in N$ such that

$$\models (\forall \mathbf{n} \in \mathbf{N}^+)(\forall \mathbf{m} \in \mathbf{N}^+)((\mathbf{n} > \mathbf{n}_0 \ \& \ \mathbf{m} > \mathbf{n}_0) \rightarrow \rho(\mathbf{s}_n,\mathbf{s}_m) < \varepsilon).$$

By the Transfer Principle, we see that $n,m \in {}^*N - N$ implies $\rho(s_n,s_m) < \varepsilon$. Since ε was arbitrary, $\rho(s_n,s_m) \approx 0$ and $s_n \approx s_m$.

Conversely, let $s_n \approx s_m$ for $m,n \in {}^*N - N$. Let $\varepsilon \in R^+$. By choosing for n_0 any infinite integer, it follows that

$$* \models (\exists \mathbf{n}_0 \in {}^*\mathbf{N}^+)(\forall \mathbf{m} \in {}^*\mathbf{N}^+)(\forall \mathbf{n} \in \mathbf{N})((\mathbf{n} > \mathbf{n}_0 \ \& \ \mathbf{m} > \mathbf{n}_0) \rightarrow \rho(\mathbf{s}_m,\mathbf{s}_n) < \varepsilon).$$

The result follows by an application of the Transfer Principle. ■

DEFINITION. X is called a *complete metric space* if every Cauchy sequence of points of the space has a limit.

THEOREM 5.14. $s_n \rightarrow x$ implies $\{s_n\}$ is a Cauchy sequence.

PROOF. Since $\{s_n | n \in N^+\}$ has a limit x, $s_\nu \approx x$ and $s_\mu \approx x$ for any $\nu,\mu \in {}^*N - N$. Hence $s_\nu \approx s_\mu$. ■

THEOREM 5.15. If a Cauchy sequence $\{s_n\}$ has a limit point, then it has a limit.

PROOF. Since $\{s_n\}$ has a limit point x, we have $s_{\nu_0} \approx x$ for some $\nu_0 \in {}^*N - N$. Since $\{s_n\}$ is a Cauchy sequence, $s_\nu \approx s_{\nu_0}$ for all $\nu \in {}^*N - N$. Hence $s_\nu \approx x$ for all $\nu \in {}^*N - N$ and $s_n \rightarrow x$. ■

COROLLARY 5.16. A compact metric space is complete.

PROOF. In a compact metric space every sequence has a limit point. ■

THEOREM 5.17. A Cauchy sequence is bounded.

PROOF. Let $\{s_n | n \in N^+\}$ be a Cauchy sequence, and let ν be an arbitrary infinite integer. As in the proof of Corollary 5.9, it suffices to show that s_ν is finite.

Consider the set:

$$A = \{n \in {}^*N | * \models \boldsymbol{\rho}(\mathbf{s}_\nu, \mathbf{s}_n) < \mathbf{1}\}.$$

Clearly, A is a definable subset of $^*N$; hence it is internal. Since $\{s_n | n \in N^+\}$ is a Cauchy sequence, $^*N - N \subseteq A$. Since $^*N - N$ is external and A is internal, $A \cap N \neq \varnothing$. Hence there is a finite n such that $* \models \rho(s_\nu, s_n) < 1$ and s_ν must be finite. ■

THEOREM 5.18. Let X be a complete metric space, and let A be a closed subset of X. Then, regarded as a subspace of X, A is also complete.

PROOF. Let $\{s_n\}$ be a Cauchy sequence of points of A. Since X is complete, $s_n \to x$ for some $x \in X$. For $n \in {}^*N - N$, $s_n \in {}^*A$ (by the Transfer Principle) and $s_n \approx x$. Since A is closed, $x \in A$. Hence A is complete. ■

THEOREM 5.19. If every finite point in X is near standard, then X is complete.

PROOF. Let $\{s_n | n \in N^+\}$ be a Cauchy sequence. It suffices to show that the sequence has a limit point. Let ν be some fixed infinite integer. By Theorem 5.17 s_ν is finite; hence $s_\nu \approx x \in X$. But then by Theorem 5.11 x is a limit point of the sequence. ■

The next theorem gives a useful nonstandard equivalent of completeness.

THEOREM 5.20. X is complete if and only if for every remote $p \in {}^*X$ there exists a real $r > 0$ such that $\rho(p,q) \geq r$ for all $q \in X$.

PROOF. Suppose that X is complete, but the conclusion fails to hold. Then for some remote $p \in {}^*X$, there is a standard sequence $\{q_n | n \in N^+\}$ such that $\rho(p,q_n) < 1/n$ for $n \in N^+$. Then

$$\rho(q_m,q_n) \leq \rho(q_m,p) + \rho(q_n,p) < (1/n) + (1/m),$$

so the sequence $\{q_n | n \in N^+\}$ is a Cauchy sequence, and thus has a limit $q \in X$. But then

$$\begin{aligned} \rho(p,q) &\leq \rho(p,q_n) + \rho(q_n,q) \\ &< (1/n) + \rho(q_n,q) \end{aligned}$$

for all $n \in N^+$, so that $\rho(p,q) \approx 0$, which contradicts the remoteness of p.

Proceeding conversely, let $\{s_n | n \in N^+\}$ be a Cauchy sequence, and let ν be some infinite integer. By Theorems 5.11 and 5.15, it suffices to show that s_ν is near standard. Suppose s_ν is remote. By hypothesis, there exists a real

$r > 0$ such that

$$\rho(s_\nu,q) \geq r \quad \text{for all} \quad q \in X.$$

In particular,

$$\rho(s_\nu,s_n) \geq r \quad \text{for all} \quad n \in N^+.$$

Since $\{s_n | n \in N^+\}$ is a Cauchy sequence, there is an $n_0 \in N^+$ such that

$$\models (\forall \mathbf{n} \in \mathbf{N}^+)(\forall \mathbf{m} \in \mathbf{N}^+)((\mathbf{m} > \mathbf{n}_0 \ \&\ \mathbf{n} > \mathbf{n}_0) \rightarrow \boldsymbol{\rho}(\mathbf{s}_m,\mathbf{s}_n) < \mathbf{r}).$$

By the Transfer Principle $\rho(s_m,s_n) < r$ for all $m,n \in {}^*N$ with $m,n > n_0$. Choosing $m = \nu$ and n any natural number $> n_0$, we get a contradiction. ■

We conclude this section with a result on internality that will be quite useful later.

THEOREM 5.21 (REMOTENESS THEOREM). Let $\{p_n | n \in {}^*N^+\}$ be an internal sequence of points of ${}^*X$. Let $r > 0$ be real. Suppose $\rho(p_j,p_k) \geq r$ for all $j \neq k$, $j,k \in N^+$. Then for some $n \in {}^*N^+$, p_n is remote.

PROOF. Suppose that p_n is near standard for all $n \in {}^*N^+$. Let $q_n = {}^\circ(p_n)$ for $n \in N^+$. Since q is a standard mapping of N^+ into X, ${}^*q$ (which, as usual, we write as q) maps ${}^*N^+$ into ${}^*X$. (Of course, the relation $q_n \approx p_n$ is only known to hold for $n \in N^+$.) Consider the internal sequence (cf. Theorem 1–8.9):

$$\{\rho(q_n,p_n) | n \in {}^*N^+\}.$$

Since $\rho(q_n,p_n) \approx 0$ for $n \in N^+$, by the Infinitesimal Prolongation Theorem there exists $\nu \in {}^*N - N$ such that $\rho(q_n,p_n) \approx 0$ for all $n < \nu$. Choose $\mu < \nu$, $\mu \in {}^*N - N$. Since p_μ is near standard, we can set $q = {}^\circ(p_\mu)$. Since $\rho(p_\mu,q_\mu) \approx 0$ and $\rho(p_\mu,q) \approx 0$, it follows that $q_\mu \approx q$. Hence q is a limit point of the sequence $\{q_n | n \in N^+\}$. Thus there is a subsequence $q_{n_j} \rightarrow q$. Since this subsequence converges, it is a Cauchy sequence. Therefore, there is a j_0 such that for $j,k > j_0$, $\rho(q_{n_j},q_{n_k}) < r/2$. Since

$$\rho(p_{n_j},p_{n_k}) \leq \rho(p_{n_j},q_{n_j}) + \rho(q_{n_j},q_{n_k}) + \rho(q_{n_k},p_{n_k}),$$

and since the first and third terms on the right are infinitesimal, it follows that $\rho(p_{n_j},p_{n_k}) < r$. This is a contradiction. ■

6. UNIFORM CONVERGENCE

In this section, we are concerned with functions f that map a metric space X into a metric space T, as well as with sequences of such functions.

DEFINITION. The sequence $\{f_n | n \in N^+\}$ *converges uniformly to f on X*, written $f_n \rightrightarrows f$, if for every real $\varepsilon > 0$ there is an $n_0 \in N^+$ such that for all $p \in X$

$$n > n_0 \quad \text{implies} \quad \rho(f_n(p),f(p)) < \varepsilon.$$

THEOREM 6.1. $f_n \rightrightarrows f$ on X if and only if $f_\nu(p) \approx f(p)$ for all $p \in {}^*X$ and $\nu \in {}^*N - N$.

PROOF. Let $f_n \rightrightarrows f$ on X, and let ε be any positive real number. Then there exists an $n_0 \in N^+$ such that

$$\models (\forall \mathbf{n} \in \mathbf{N}^+)(\forall \mathbf{p} \in \mathbf{X})(\mathbf{n} > \mathbf{n}_0 \rightarrow \boldsymbol{\rho}(\mathbf{f}_n(\mathbf{p}), \mathbf{f}(\mathbf{p})) < \varepsilon).$$

By the Transfer Principle, it follows that for any $\nu \in {}^*N - N$ and $p \in {}^*X$,

$$\rho(f_\nu(p), f(p) < \varepsilon.$$

Since this holds for all real $\varepsilon > 0$,

$$\rho(f_\nu(p), f(p)) \approx 0,$$

That is,

$$f_\nu(p) \approx f(p).$$

Conversely, assume that if ν is infinite and $p \in {}^*X$, then $\rho(f_\nu(p), f(p)) \approx 0$. Hence if ε is any fixed positive real number,

$$* \models (\exists \mathbf{n}_0 \in {}^*\mathbf{N}^+)(\forall \mathbf{n} \in {}^*\mathbf{N}^+)(\forall \mathbf{p} \in {}^*\mathbf{X})(\mathbf{n} > \mathbf{n}_0 \rightarrow \boldsymbol{\rho}(\mathbf{f}_n(\mathbf{p}), \mathbf{f}(\mathbf{p})) < \varepsilon)).$$

Applying the Transfer Principle, we see that $f_n \rightrightarrows f$ on S. ■

THEOREM 6.2. Let $f_n \rightrightarrows f$ on X where each f_n is continuous on X. Then f is continuous on X.

PROOF. Let $q \in X$, $p \in {}^*X$, $p \approx q$. We need to show that $f(p) \approx f(q)$. Since for n finite each f_n is continuous, the internal sequence

$$\{\rho(f_n(p), f_n(q)) | n \in {}^*N^+\}$$

of hyperreal numbers is infinitesimal for finite n. Hence by the Infinitesimal Prolongation Theorem, there exists a $\nu \in {}^*N - N$ such that

$$\rho(f_n(p), f_n(q)) \approx 0$$

for all $n < \nu$. Let $\mu \in {}^*N - N$ where $\mu < \nu$. Then

$$f_\mu(p) \approx f_\mu(q).$$

By hypothesis, $f(p) \approx f_\mu(p)$, $f_\mu(q) \approx f(q)$. Hence $f(p) \approx f(q)$. ■

THEOREM 6.3 (DINI). Let $\{f_n | n \in N^+\}$ be a sequence of continuous functions from the compact space X into R. Let $f_{n+1}(p) \leq f_n(p)$ for $n \in N^+$ and $p \in X$. Finally, let $f_n(p) \rightarrow f(p)$ for each point $p \in X$ where f is continuous on X. Then $f_n \rightrightarrows f$ on X.

PROOF. Without loss of generality we can assume that $f(p) = 0$ for $p \in X$. (For, otherwise, we could apply this special case to the sequence $\{f_n - f | n \in N^+\}$.)

Now, by hypothesis, if $n \geq m$, $n,m \in N^+$, then $0 \leq f_n(p) \leq f_m(p)$ for all $p \in X$. Let $\nu \in {}^*N - N$ and $p \in {}^*X$. It suffices to show that $f_\nu(p) \approx 0$. By the Transfer Principle, $0 \leq f_\nu(p) \leq f_n(p)$ for all $n \in N^+$. Since X is compact, $p \approx q$ for some $q \in X$. $f_n(p) \approx f_n(q)$ for each $n \in N^+$ because each f_n is continuous. Since $f_n(q) \in R$, $f_n(q) = {}^\circ(f_n(p))$. Since $f_n(p)$ is finite, $f_\nu(p)$ is also finite and hence near standard. It follows that

$$0 \leq {}^\circ(f_\nu(p)) \leq {}^\circ(f_n(p)) = f_n(q)$$

for all $n \in N^+$. Since $f_n(q) \to 0$, we have ${}^\circ(f_\nu(p)) = 0$, that is, $f_\nu(p) \approx 0$. ■

7. UNIFORM CONTINUITY AND EQUICONTINUITY

We continue to consider maps from a metric space X into a metric space T.

DEFINITION. f is *uniformly continuous on* X if for every real $\varepsilon > 0$, there is a real $\delta > 0$ such that for all $p,q \in X$,

$$\rho(p,q) < \delta \quad \text{implies} \quad \rho(f(p),f(q)) < \varepsilon.$$

In considering maps from $^*X$ into $^*T$, we use several different notions of continuity.

DEFINITION. Let f map $^*X$ into $^*T$, where $q \in X^*$. Then f is called *microcontinuous* at q if $p \approx q$ implies $f(p) \approx f(q)$ for all $p \in {}^*X$. f is microcontinuous on $^*X$ if f is microcontinuous at all points $q \in {}^*X$.

THEOREM 7.1. f is uniformly continuous on X if and only if f is microcontinuous on $^*X$.

PROOF. Let f be uniformly continuous on X. Then for each real $\varepsilon > 0$ there exists a real $\delta > 0$ such that

$$\models (\forall \mathbf{p} \in \mathbf{X})(\forall \mathbf{q} \in \mathbf{X})(\boldsymbol{\rho}(\mathbf{p},\mathbf{q}) < \boldsymbol{\delta} \rightarrow \boldsymbol{\rho}(\mathbf{f}(\mathbf{p}),\mathbf{f}(\mathbf{q})) < \boldsymbol{\varepsilon}).$$

By the Transfer Principle, it follows that for $p,q \in {}^*X$

$$\rho(p,q) < \delta \quad \text{implies} \quad \rho(f(p),f(q)) < \varepsilon.$$

Thus $p \approx q$ implies that $\rho(f(p),f(q))$ is less than every real $\varepsilon > 0$ so that $\rho(f(p),f(q)) \approx 0$. That is, f is microcontinuous on $^*X$.

Conversely, let f be microcontinuous on $^*X$. Let ε be some positive real number. Then if δ is any positive infinitesimal and $p,q \in {}^*X$ are such that $\rho(p,q) < \delta$, then $\rho(f(p),f(q)) < \varepsilon$. Hence

$$* \models (\exists \boldsymbol{\delta} \in {}^*\mathbf{R}^+)(\forall \mathbf{p} \in \mathbf{X})(\forall \mathbf{q} \in \mathbf{X})(\boldsymbol{\rho}(\mathbf{p},\mathbf{q}) < \boldsymbol{\delta} \rightarrow \boldsymbol{\rho}(\mathbf{f}(\mathbf{p}),\mathbf{f}(\mathbf{q})) < \boldsymbol{\varepsilon}).$$

The result follows from the Transfer Principle. ■

THEOREM 7.2. If f is continuous on the compact space X, then f is uniformly continuous.

PROOF. By the previous theorem, it suffices to show f is microcontinuous on $*X$. Let $q,p \in *X$ where $q \approx p$. We have to show that $f(q) \approx f(p)$. Since X is compact, $q \approx s$ for some $s \in X$. Hence also $p \approx s$. Then $f(q) \approx f(s)$ and $f(p) \approx f(s)$ because f is continuous on X. Hence $f(q) \approx f(p)$. ■

DEFINITION. The sequence $\{f_n | n \in N^+\}$, where each f_n maps X into T, is called *equicontinuous* if for every real $\varepsilon > 0$ there is a real $\delta > 0$ such that for all $p,q \in X$ and all $n \in N^+$,

$$\rho(p,q) < \delta \quad \text{implies} \quad \rho(f_n(p),f_n(q)) < \varepsilon.$$

THEOREM 7.3. $\{f_n | n \in N^+\}$ is equicontinuous on X if and only if f_n is microcontinuous on $*X$ for each $n \in *N^+$.

PROOF. Let $\{f_n\}$ be equicontinuous on X, and let ε be a positive real number. Then there exists a real $\delta > 0$ such that

$$\models (\forall \mathbf{n} \in \mathbf{N}^+)(\forall \mathbf{p} \in \mathbf{X})(\forall \mathbf{q} \in \mathbf{X})(\boldsymbol{\rho}(\mathbf{p},\mathbf{q}) < \boldsymbol{\delta} \rightarrow \boldsymbol{\rho}(\mathbf{f}_n(\mathbf{p}),\mathbf{f}_n(\mathbf{q})) < \boldsymbol{\varepsilon}).$$

Using the Transfer Principle, we see that

$$\rho(p,q) < \delta \quad \text{implies} \quad \rho(f_n(p),f_n(q)) < \varepsilon$$

for $p,q \in *X$ and all $n \in *N^+$. Since ε is arbitrary, it follows that $p \approx q$ implies $f_n(p) \approx f_n(q)$. Hence f_n is microcontinuous for $n \in *N$.

Conversely, let f_n be microcontinuous on $*X$ for each $n \in *N^+$. Let ε be any positive real number. Then

$$* \models (\exists \boldsymbol{\delta} \in *\mathbf{R})(\forall \mathbf{p} \in *\mathbf{X})(\forall \mathbf{q} \in *\mathbf{X})(\boldsymbol{\rho}(\mathbf{p},\mathbf{q}) < \boldsymbol{\delta} \rightarrow \boldsymbol{\rho}(\mathbf{f}_n(\mathbf{p}),\mathbf{f}_n(\mathbf{q})) < \boldsymbol{\varepsilon}).$$

The result follows as usual by an application of the Transfer Principle. ■

THEOREM 7.4. If $f_n(p) \rightarrow f(p)$ for every $p \in X$ and $\{f_n | n \in N^+\}$ is equicontinuous, then f is uniformly continuous on X.

PROOF. By hypothesis,

$$\models (\forall \boldsymbol{\varepsilon} \in \mathbf{R}^+)(\forall \mathbf{u} \in \mathbf{X})(\exists \mathbf{m} \in \mathbf{N}^+)(\forall \mathbf{n} \in \mathbf{N}^+)(\mathbf{n} > \mathbf{m} \rightarrow \boldsymbol{\rho}(\mathbf{f}_n(\mathbf{u})\mathbf{f},(\mathbf{u})) < \boldsymbol{\varepsilon}).$$

Letting ε be a positive infinitesimal, and choosing $p,q \in *X$, we have by the Transfer Principle that there exist $v_p,v_q \in *N^+$ such that

$$n > v_p \quad \text{implies} \quad \rho(f_n(p),f(p)) < \varepsilon,$$
$$n > v_q \quad \text{implies} \quad \rho(f_n(q),f(q)) < \varepsilon.$$

Thus if μ is some infinite integer $> \nu_p, \nu_q$, and if $p \approx q$, then

$$f(p) \approx f_\mu(p) \approx f_\mu(q) \approx f(q).$$

Thus f is uniformly continuous on S. ■

THEOREM 7.5. Let $f_n(p) \to f(p)$ for each $p \in X$, and let X be a compact space. Let f_n be continuous on X for each $n \in N^+$. Then $f_n \rightrightarrows f$ on X if and only if $\{f_n | n \in N^+\}$ is equicontinuous.

PROOF. Let $f_n \rightrightarrows f$ on X; thus f is continuous on X. Since X is compact, f and all the f_n, $n \in N^+$ are uniformly continuous on X. Hence f and all the f_n, $n \in N^+$ are microcontinuous on $^*X$. Now, let $\nu \in {}^*N - N$. Then, if $p,q \in {}^*X$ and $p \approx q$, we have

$$f_\nu(p) \approx f(p) \approx f(q) \approx f_\nu(q).$$

Thus f_n is microcontinuous for $n \in {}^*N - N$. Since we already know this for $n \in N^+$, we have that $\{f_n | n \in N^+\}$ is equicontinuous.

Conversely, let $\{f_n | n \in N^+\}$ be equicontinuous so that f_n is microcontinuous on $^*X$ for each $n \in {}^*N^+$. Let $q \in {}^*X$. Since X is compact, $q \approx p \in X$ and $f_n(q) \approx f_n(p)$ for all $n \in {}^*N^+$. For ν infinite, moreover, $f_\nu(p) \approx f(p)$. By Theorem 7.4, f is (uniformly) continuous on X, so $f(q) \approx f(p)$. Thus for $\nu \in {}^*N - N$,

$$f_\nu(q) \approx f_\nu(p) \approx f(p) \approx f(q),$$

and so $f_n \rightrightarrows f$ on X. ■

It is revealing to compare microcontinuity with two related notions. Thus

DEFINITION. Let f map $^*X$ into $^*T$, and let $q \in {}^*X$. Then f is *εδ-continuous* at q, if for any *real* $\varepsilon > 0$ there is a *real* $\delta > 0$ such that for all $p \in {}^*X$,

$$\rho(p,q) < \delta \quad \text{implies} \quad \rho(f(p),f(q)) < \varepsilon. \tag{‡}$$

f is *$*$-continuous at q* if for every *hyperreal* $\varepsilon > 0$, there is a *hyperreal* $\delta > 0$ such that for all $p \in {}^*X$ (‡) holds.

LEMMA 1. If f is continuous on X, then f is $*$-continuous at all points $q \in {}^*X$.

PROOF. We have

$$\models (\forall q \in X)(\forall \varepsilon \in R^+)(\exists \delta \in R^+)(\forall x \in X)(\rho(q,x) < \delta \to \rho(f(q)f(x)) < \varepsilon).$$

Using the Transfer Principle, we see that f is $*$-continuous at each $q \in {}^*X$. ■

LEMMA 2. If f is $\varepsilon\delta$-continuous at q, then f is microcontinuous at q.

PROOF. Let ε be any positive real number. By hypothesis, we can find a real number $\delta > 0$ such that for all $p \in {}^*X$

$$\rho(p,q) < \delta \quad \text{implies} \quad \rho(f(p),f(q)) < \varepsilon.$$

Thus if $p \approx q$, then $\rho(f(p),f(q)) < \varepsilon$. Since ε was an arbitrary positive real number, $f(p) \approx f(q)$. ■

To see that the converse of this lemma is false, note the

EXAMPLE 1. Let f be defined on $^*R$ as follows:

$$f(x) = \begin{cases} 0 & \text{if } x \approx 0, \\ 1 & \text{otherwise.} \end{cases}$$

f is clearly microcontinuous. However, f is not $\varepsilon\delta$-continuous. For given any real $\delta > 0$, we have

$$\rho\left(\frac{\delta}{2}, 0\right) < \delta, \quad \text{but} \quad \rho\left(f\left(\frac{\delta}{2}\right), f(0)\right) = 1.$$

We next give an example of a function that is $*$-continuous, but not microcontinuous.

EXAMPLE 2. Let $f(x) = x^2$ for $x \in R$. By the Transfer Principle, $f(x) = x^2$ for all $x \in {}^*R$. Since f is continuous on R, by Lemma 1, f is a $*$-continuous function on each point of $^*R$. However, f is not microcontinuous on $^*R$. In fact, let $\nu \in {}^*N - N$. Then $\nu + (1/\nu) \approx \nu$, but

$$\left(\nu + \frac{1}{\nu}\right)^2 = \nu^2 + 2 + \frac{1}{\nu^2} \approx \nu^2 + 2 \not\approx \nu^2.$$

Finally, we examine a function that is microcontinuous (and internal) but not $*$-continuous.

EXAMPLE 3. Let $\alpha > 0$ where $\alpha \approx 0$. Let

$$f(x) = \begin{cases} \alpha \sin \dfrac{1}{x} & \text{for} \quad x \neq 0 \text{ and } x \in {}^*R, \\ 0 & \text{if} \quad x = 0. \end{cases}$$

At $x = 0$, f is microcontinuous, but not $*$-continuous. To verify microcontinuity, let $x \approx 0$, $x \neq 0$; then using the Transfer Principle,

$$|f(x) - f(0)| = \left|\alpha \sin \frac{1}{x}\right| \leq |\alpha| \approx 0.$$

To see that f is not $*$-continuous at 0, let $\varepsilon = \alpha/2$. Now for $x = 1/(2n\pi + \pi/2)$, $n \in {}^*N$, $\sin(1/x) = 1$. Hence, for any positive hyperreal δ, we can find x such that $0 < x < \delta$ and

$$|f(x) - f(0)| = \alpha \sin \frac{1}{x} = \alpha > \varepsilon.$$

Although microcontinuity does not imply $\varepsilon\delta$-continuity in general, it does so for internal functions.

LEMMA 3. If f is internal and microcontinuous at q, then f is $\varepsilon\delta$-continuous at q.

PROOF. Let ε be any positive real number. Let

$$A = \{n \in {}^*N^+ | * \models (\forall \mathbf{p} \in {}^*\mathbf{X})(\boldsymbol{\rho}(\mathbf{p},\mathbf{q}) < \mathbf{1/n} \rightarrow \boldsymbol{\rho}(\mathbf{f}(\mathbf{p}),\mathbf{f}(\mathbf{q})) < \boldsymbol{\varepsilon})\}.$$

A is clearly a definable subset of ${}^*N$ and thus is internal. On the other hand, by the microcontinuity of f at q, ${}^*N - N \subseteq A$. Since A is internal and ${}^*N - N$ is external, $A \cap N \neq \varnothing$. Let $\delta = 1/k$ for some $k \in A \cap N$. Then for any $p \in {}^*X$,

$$\rho(p,q) < \delta \quad \text{implies} \quad \rho(f(p),f(q)) < \varepsilon,$$

that is, f is $\varepsilon\delta$-continuous at q. ■

These considerations lead to the following important

THEOREM 7.6 (APPROXIMATION THEOREM). Let X be a compact space. Let f be microcontinuous on ${}^*X$ and internal, and let $f(p)$ be near standard for all $p \in X$. Let $F(p) = {}^\circ(f(p))$ for all points $p \in X$. Then F is continuous on X, and $F(p) \approx f(p)$ for all $p \in {}^*X$.

PROOF. Let $p \in X$. We show that F is continuous at p. Since f is microcontinuous and internal, f is also $\varepsilon\delta$-continuous. Let ε be any positive real number. By $\varepsilon\delta$-continuity there exists a real $\delta > 0$ such that for all $q \in {}^*X$,

$$\rho(q,p) < \delta \quad \text{implies} \quad \rho(f(q),f(p)) < \frac{\varepsilon}{2}.$$

So let q be any point of X such that $\rho(q,p) < \delta$. By the way F is defined, $f(q) \approx F(q)$, $f(p) \approx F(p)$. Hence

$$\rho(F(q),F(p)) \leq \rho(F(q),f(q)) + \rho(f(q),f(p)) + \rho(f(p),F(p)) < \varepsilon.$$

Thus F is continuous at p. Since p was any point of X, F is continuous on X.

Now, let $p \in {}^*X$. Since X is compact, $p \approx q \in X$. Since f is microcontinuous, $f(p) \approx f(q)$. By the definition of $F, f(q) \approx F(q)$. Finally, since (as we have just seen) F is continuous, $F(p) \approx F(q)$. Hence $F(p) \approx f(p)$. ■

We apply this Approximation Theorem to prove the famous

THEOREM 7.7 (ARZELÀ–ASCOLI THEOREM). An equicontinuous sequence of functions from a compact space X to a compact space T has a uniformly convergent subsequence.

PROOF. Let $\{f_n | n \in N^+\}$ be the given equicontinuous family. Let v be some fixed infinite integer. Then by Theorem 7.3, f_v is microcontinuous on $^*X$. Since f_v is clearly internal, the Approximation Theorem can be applied. That is, we set $F(p) = {}^\circ(f_v(p))$ for $p \in X$ and conclude that $f_v(p) \approx F(p)$ for all $p \in {}^*X$. Thus for any fixed $k,m \in N^+$, we have

$$* \models (\exists \mathbf{q} \in {}^*\mathbf{N})(\mathbf{q} > \mathbf{m} \;\&\; (\forall \mathbf{x} \in {}^*\mathbf{X})(\boldsymbol{\rho}(\mathbf{F}(\mathbf{x}),\mathbf{f}_q(\mathbf{x})) < \mathbf{1/k})).$$

Using the Transfer Principle, we conclude that there is a map q on $N \times N$ into N such that $q(m,k) > m$ and for all $x \in X$, $\rho(F(x), f_{q(m,k)}(x)) < 1/k$. Letting $n_0 = 1, n_{i+1} = q(n_i, i + 1)$ we see that $n_{i+1} > n_i$ and that $\rho(F(x), f_{n_i}(x)) < 1/i$ for $i \geq 1$. Thus $f_{n_i} \rightrightarrows F$. ■

8. COMPACT MAPPINGS

Let X and T be metric spaces, and let f map X into T.

DEFINITION. f is *compact* if for every bounded subset B of $X, f[B] \subseteq K$ for some compact subset K of T.

A nonstandard criterion for a function to be compact is given in the following:

THEOREM 8.1. f is compact if and only if f maps finite points of $^*X$ into near-standard points of $^*T$.

PROOF. Let f be compact, and let p be a finite point of $^*X$. We have to show that $f(p)$ is near standard. Since p is finite, there is a $q \in X$ such that $\rho(p,q) \leq r$ for some $r \in R^+$. Let $B = \{x \in X | \rho(x,q) \leq r\}$. Since $B \subseteq B_q(r + 1)$, B is bounded. Hence $f[B] \subseteq K$ for some compact set K. Since $p \in {}^*B$, $f(p) \in f[{}^*B] = {}^*(f[B])$. Hence $f(p) \in {}^*K$. Since K is compact, $f(p)$ is near standard.

Conversely, let f map finite points of $^*X$ into near-standard points of $^*T$, and let B be a bounded subset of X. We need to show that $f[B] \subseteq K$ for some compact set K. We let K consist of all points $q \in T$ such that $q \approx p$ for some $p \in {}^*(f[B])$. Clearly, $f[B] \subseteq K$. It remains to show that K is compact.

Now if $q \in K$ and ε is a positive real number, we have

$$* \models (\exists \mathbf{p} \in {}^*(\mathbf{f}[\mathbf{B}]))\, \boldsymbol{\rho}(\mathbf{p},\mathbf{q}) < \boldsymbol{\varepsilon}.$$

By the Transfer Principle, it follows that for every $q \in K$ and $\varepsilon \in R^+$ there exists a $p \in f[B]$ such that $\rho(q,p) < \varepsilon$. That is,

$$\models (\forall \mathbf{q} \in \mathbf{K})(\forall \varepsilon \in \mathbf{R}^+)(\exists \mathbf{p} \in \mathbf{f}[\mathbf{B}])\, \rho(\mathbf{q},\mathbf{p}) < \varepsilon.$$

Again applying the Transfer Principle, we conclude that if $q \in {}^*K$ and ε is a positive infinitesimal, there must be a $p \in {}^*(f[B]) = f[{}^*B]$ such that $\rho(q,p) < \varepsilon$, that is, $q \approx p$. Now since B is bounded, the points of ${}^*B$ are finite, and thus by hypothesis p is near standard, say $p \approx s$. By the definition of the set K, $s \in K$ and $q \approx p \approx s$. Thus any point of ${}^*K$ is infinitesimally near a point of K; that is, K is compact. ■

We apply this in the following:

THEOREM 8.2. Let $\{f_n | n \in N^+\}$ be a sequence of compact maps from X into T where T is a complete metric space. Let $f_n \rightrightarrows f$ on X. Then f is compact.

PROOF. Let $p \in {}^*X$ where p is finite. By Theorem 8.1, we need only show that $f(p)$ is near standard. Suppose instead that $f(p)$ were remote. Then, by the Remoteness Theorem (5.21), there exists $r \in R^+$ such that $\rho(f(p),q) \geq r$ for all $q \in T$. But there exists an $n_0 \in N$ such that $\rho(f_{n_0}(s), f(s)) < r/2$ for all $s \in X$. By the Transfer Principle, we have that $\rho(f_{n_0}(p), f(p)) < r/2$. Since f_{n_0} is compact and p is finite, $f_{n_0}(p)$ is near standard. Hence $f_{n_0}(p) \approx q$ for some $q \in T$. Then,

$$\rho(f(p),q) \leq \rho(f(p), f_{n_0}(p)) + \rho(f_{n_0}(p),q) < r.$$

This is a contradiction. ■

4

NORMED LINEAR SPACES

1. LINEAR SPACES

In this chapter, we deal with linear spaces $\mathcal{N}$ over a field D. For our purposes, we assume that $D, \mathcal{N} \subseteq S$. Then, since the axioms for a linear space are expressible as the truth in U of sentences of $\mathcal{L}$, we have, by the Transfer Principle, that $^*\mathcal{N}$ is a linear space over the field $^*D$.

In fact, we work with just two fields: the field of real numbers R and the complex number field, which we denote by C. Since $z \in C$ if and only if $z = x + iy$ with $x,y \in R$, we have that

$$^*C = \{x + iy | x,y \in {}^*R\}.$$

For $z = x + iy \in {}^*C$, $|z|$ is the hyperreal number $(x^2 + y^2)^{1/2}$. Thus if z is finite, so are x and y, and we may write

$$^\circ z = {}^\circ x + i({}^\circ y).$$

It is trivial to verify that $^\circ$ as thus defined is a homomorphism of $^*C$ into C.

DEFINITION. A normed linear space $\mathcal{N}$ over D is a linear space over D together with a map $\|\cdots\|$ of $\mathcal{N}$ into R such that for all $x,y \in \mathcal{N}$ and $\alpha \in D$:

(1) $\|x\| \geq 0$,
(2) $\|x\| = 0$ if and only if $x = 0$,
(3) $\|\alpha x\| = |\alpha| \|x\|$,
(4) $\|x + y\| \leq \|x\| + \|y\|$ (the triangle inequality).

$\|x\|$ is called the *norm* of x.

It is clear that the function $\rho(x,y) = \|x - y\|$ is a metric on a normed linear space so that the results of Chapter 3 can be applied. If, in particular, $\mathcal{N}$ is a *complete* metric space, then it is called a *Banach space*.

In the nonstandard universe, $\|x\| \in {}^*R$ is defined for all $x \in {}^*\mathcal{N}$, and by the Transfer Principle, (1) to (4) hold for $x,y \in {}^*\mathcal{N}$, $\alpha \in {}^*D$.

It should be noted that for either $D = R$ or $D = C$, D itself is a normed linear space over D using the field operations and $\|x\| = |x|$.

$x \in {}^*\mathcal{N}$ is called *infinitesimal* if $x \approx 0$. Thus

THEOREM 1.1. An element $x \in {}^*\mathcal{N}$ is infinitesimal if and only if $\|x\| \approx 0$.

PROOF.

$$\begin{aligned} x \approx 0 \quad &\text{if and only if} \quad \rho(x,0) \approx 0 \\ &\text{if and only if} \quad \|x\| \approx 0. \quad \blacksquare \end{aligned}$$

THEOREM 1.2. If $x,y \in {}^*\mathcal{N}$ and $x \approx y$, then $\|x\| \approx \|y\|$.

PROOF. Suppose (without loss of generality) that $\|x\| \le \|y\|$. Then $\|y\| = \|x + (y - x)\| \le \|x\| + \|y - x\|$. So,

$$0 \le \|y\| - \|x\| \le \|y - x\| \approx 0.$$

It follows that $\|y\| - \|x\| \approx 0$, that is, $\|x\| \approx \|y\|$. $\blacksquare$

THEOREM 1.3. $x \in {}^*\mathcal{N}$ is a finite point (in the sense of Section 5 of Chapter 3) if and only if $\|x\|$ is a finite hyperreal number.

PROOF. If $\|x\|$ is finite, then $\|x - 0\|$ is finite so that x is a finite point. Conversely, if x is finite, then $\|x - y\|$ is finite for some $y \in \mathcal{N}$. Thus

$$\|x\| = \|(x - y) + y\| \le \|x - y\| + \|y\|$$

is also finite. $\blacksquare$

If $x_1,x_2,y_1,y_2 \in {}^*\mathcal{N}$, then as is readily verified, the relations $x_1 \approx x_2$, $y_1 \approx y_2$ imply $x_1 + y_1 \approx x_2 + y_2$. Also, if α is a *finite* element of ${}^*D$, then $\alpha x_1 \approx \alpha x_2$.

If $x \in {}^*\mathcal{N}$ and $x \approx y \in \mathcal{N}$, we write $y = {}^\circ x$ in agreement with the notation of Section 5 of Chapter 3.

THEOREM 1.4. $B \subseteq \mathcal{N}$ is a bounded set if and only if $\|x\| \le M$ for some $M \in R$ and all $x \in B$.

PROOF. If B is bounded, then by definition, there is an $x_0 \in \mathcal{N}$ and an $M_0 \in R$ such that for all $x \in B$, $\|x - x_0\| \le M_0$. Then

$$\|x\| \le \|x - x_0\| + \|x_0\| \le M_0 + \|x_0\|$$

for $x \in B$. The converse is trivial. $\blacksquare$

DEFINITION. Let $\mathcal{N}$ and $\mathcal{M}$ be normed linear spaces over the same field D. A mapping T from $\mathcal{N}$ into $\mathcal{M}$ is called a *linear operator* if

(1) $T(x + y) = T(x) + T(y)$ for all $x,y \in \mathcal{N}$,
(2) $T(\alpha x) = \alpha T(x)$ for $x \in \mathcal{N}$, $\alpha \in D$.

If $\mathcal{M} = D$, then T is called a *linear functional.*

We often write Tx for $T(x)$. Of course, $^*T$ (which we ordinarily write as T) maps $^*\mathcal{N}$ into $^*\mathcal{M}$, and satisfies (1) and (2).

THEOREM 1.5. Let T be a linear operator on $\mathcal{N}$ that is continuous at some $x_0 \in \mathcal{N}$. Then T is uniformly continuous on $\mathcal{N}$.

PROOF. By Theorem 3–7.1, we need only show that T is microcontinuous on $^*\mathcal{N}$. So let $x,y \in {}^*\mathcal{N}$, $x \approx y$. We have $x_0 = x_0 + 0 \approx x_0 + (x - y)$. Then

$$T(x_0) \approx T(x_0 + x - y) = T(x_0) + T(x) - T(y).$$

Thus $T(x) \approx T(y)$. ■

DEFINITION. A linear operator T is *bounded* if for some $M \in R$,

$$\|Tx\| \leq M\|x\| \quad \text{for all} \quad x \in \mathcal{N}.$$

For the remainder of this section, the letter T *always stands for a linear operator.*

THEOREM 1.6. T is bounded if and only if $\{Tx | x \in \mathcal{N}, \|x\| = 1\}$ is a bounded set.

PROOF. Let $A = \{Tx | x \in \mathcal{N}, \|x\| = 1\}$. If T is bounded, say $\|Tx\| \leq M\|x\|$ for all $x \in \mathcal{N}$, then for each $y \in A$ we have $y = Tx$, $\|x\| = 1$, so that $\|y\| = \|Tx\| \leq M\|x\| = M$; hence A is bounded.

Conversely, if A is bounded, we have $\|y\| \leq M$ for some $M \in R^+$, and all $y \in A$. We claim that $\|Tx\| \leq M\|x\|$ for all $x \in \mathcal{N}$. This is obvious if $x = 0$. Otherwise, let $u = x/\|x\|$ and $y = Tu = Tx/\|x\|$. Since $\|u\| = 1$, $y \in A$. Hence $\|y\| \leq M$, that is, $\|Tx\| \leq M\|x\|$. ■

In the last part of this argument M could have been any upper bound of A. Hence the argument shows that if we define for any bounded T,

$$\|T\| = \sup_{\|x\| = 1} \|Tx\|,$$

then we have the

COROLLARY 1.7. For a bounded linear operator T, $\|Tx\| \leq \|T\| \cdot \|x\|$.

Next we discuss nonstandard criteria for a linear operator to be bounded. In these considerations we write $A = \{Tx | x \in \mathcal{N}, \|x\| = 1\}$, so that T and A are either both bounded or both unbounded. Let $B = \{x \in \mathcal{N} \mid \|x\| = 1\}$. Then $A = T[B]$, so by Theorem 1–7.13, $^*A = T[^*B]$. Hence

$$^*A = \{Tx | x \in {}^*\mathcal{N}, \|x\| = 1\}. \tag{‡}$$

THEOREM 1.8. T is bounded if and only if T maps finite points into finite points.

PROOF. If T is bounded, then $\|Tx\| \leq \|T\| \cdot \|x\|$, which by Theorem 1.3 is finite for all finite x. Conversely, let T map finite points into finite points. Now for $x \in {}^*B$, $\|x\| = 1$, so x is finite; by hypothesis, Tx is also finite. So, by (‡) every point in ${}^*A$ is finite, that is by Theorem 3–5.4, A is bounded, and, therefore, by Theorem 1.6, T is bounded. ■

THEOREM 1.9. T is bounded if and only if T is continuous on $\mathcal{N}$.

PROOF. Let T be bounded, say $\|Tx\| \leq M\|x\|$. By Theorem 1.5, it suffices to show that T is continuous at 0. But for $x \approx 0$, $\|Tx\| \leq M \cdot \|x\| \approx 0$.

Conversely, let T be continuous on $\mathcal{N}$, but suppose that T is not bounded. Then by Theorem 1.6, A is not bounded, that is, ${}^*A$ contains an infinite element. Thus by (‡), for some $x \in {}^*\mathcal{N}$, $\|x\| = 1$, but Tx is infinite. Let $y = x/\|Tx\|$. Then $\|y\| = 1/\|Tx\|$, that is, $y \approx 0$. But $\|Ty\| = 1$. This contradicts the continuity of T at 0. ■

THEOREM 1.10. T is bounded if and only if T maps near-standard points into near-standard points.

PROOF. Let T be bounded, and let $x \approx s$ for $s \in \mathcal{N}$. Since T is continuous, $Tx \approx Ts$; thus Tx is near standard.

Conversely, let T map near-standard points into near-standard points, but suppose that A is not bounded. Thus by (‡) for some $x \in {}^*\mathcal{N}$ with $\|x\| = 1$, $\|Tx\|$ is infinite. Let $y = x/(\|Tx\|)^{1/2}$. Since $\|x\| = 1$ and $\|Tx\|$ is infinite, $y \approx 0$, so that y is near standard. But

$$\|Ty\| = \|Tx\|/(\|Tx\|)^{1/2} = (\|Tx\|)^{1/2},$$

which is infinite, so that Ty is not near standard, which contradicts the hypothesis. ■

If T_1, T_2 are linear operators from normed linear space $\mathcal{N}$ into normed linear space $\mathcal{M}$ and $\alpha \in D$, it is natural to write:

$$(T_1 + T_2)x = T_1x + T_2x,$$
$$(\alpha T_1)x = \alpha(T_1x).$$

The familiar fact that $T_1 + T_2$ and αT_1 as thus defined, are themselves linear operators is readily verified. Of course, $T_1 - T_2$ simply means $T_1 + (-1)T_2$, so that $(T_1 - T_2)x = T_1x - T_2x$.

We have

THEOREM 1.11. Let T_1, T_2 be bounded linear operators from $\mathcal{N}$ into $\mathcal{N}$, and let $\alpha \in D$. Then $T_1 + T_2$, αT_1, and T_1T_2 are bounded linear operators, and

$$\|T_1 + T_2\| \leq \|T_1\| + \|T_2\|,$$
$$\|\alpha T_1\| = |\alpha| \cdot \|T_1\|,$$
$$\|T_1 \cdot T_2\| \leq \|T_1\| \cdot \|T_2\|.$$

PROOF. We have

$$\begin{aligned}\|(T_1 + T_2)x\| &\le \|T_1x\| + \|T_2x\| \\ &\le \|T_1\| \cdot \|x\| + \|T_2\| \cdot \|x\| \\ &= (\|T_1\| + \|T_2\|)\|x\|.\end{aligned}$$

Then $T_1 + T_2$ is bounded, and $\|T_1 + T_2\| \le \|T_1\| + \|T_2\|$.

$$\|\alpha T_1 x\| = |\alpha| \cdot \|T_1 x\|,$$

so that αT_1 is bounded and $\|\alpha T_1\| \le |\alpha| \cdot \|T_1\|$. Moreover, $\|\alpha T_1\| = \sup_{\|x\|=1} \|\alpha T_1 x\| = |\alpha| \cdot \sup_{\|x\|=1} \|T_1 x\| = |\alpha| \cdot \|T_1\|$.

Finally,

$$\begin{aligned}\|(T_1T_2)x\| &= \|T_1(T_2x)\| \\ &\le \|T_1\| \cdot \|T_2x\| \\ &\le \|T_1\| \cdot \|T_2\| \cdot \|x\|.\end{aligned}$$

Hence T_1T_2 is bounded, and $\|T_1T_2\| \le \|T_1\| \cdot \|T_2\|$. ■

THEOREM 1.12. Let $B = \{x \mid \|x\| \le 1\}$. Let T, T_n be bounded linear operators, and let $\|T_n - T\| \to 0$. Then, $T_n \rightrightarrows T$ on B.

PROOF.

$$\|T_nx - Tx\| = \|(T_n - T)x\| \le \|T_n - T\| \cdot \|x\| \le \|T_n - T\|$$

for $x \in B$. The result follows at once. ■

THEOREM 1.13. If T is a bounded linear operator on $\mathcal{N}$ and

$$\mathcal{M} = \{x \in \mathcal{N} \mid Tx = 0\},$$

then $\mathcal{M}$ is a *closed* linear subspace of $\mathcal{N}$.

PROOF. Let $x,y \in \mathcal{M}$, $\lambda \in D$. Then

$$\begin{aligned}T(x + y) &= Tx + Ty = 0, \\ T(\lambda x) &= \lambda Tx = 0.\end{aligned}$$

Therefore, $x + y$ and λx are in $\mathcal{M}$, so that $\mathcal{M}$ is a linear subspace of $\mathcal{N}$.

To see that $\mathcal{M}$ is closed, let $x \in {}^*\mathcal{M}$, $x \approx y \in \mathcal{N}$. Then, $Tx = 0$ and $Tx \approx Ty$. Since $Ty \in \mathcal{N}$, $Ty = 0$. So, $y \in \mathcal{M}$. ■

We mention in passing the special case of Euclidean n-space, that is, $\mathcal{N} = R^n$, which is a linear space over R under the usual operations of component-wise addition and scalar multiplication. The Euclidean norm is defined by

$$\|x\| = (a_1^2 + \cdots + a_n^2)^{1/2} \quad \text{for} \quad x = \langle a_1, \ldots, a_n \rangle.$$

The fact that the Euclidean norm satisfies the triangle inequality is an elementary inequality (Minkowski's inequality).

THEOREM 1.14. Every finite point of $^*R^n$ is near standard.

PROOF. If $\|x\| = (a_1^2 + \cdots + a_n^2)^{1/2}$ is finite, then so are $|a_1|, \ldots, |a_n|$. Hence $a_1 \ldots, a_n$ are finite hyperreals, and so we may set $b_i = {}^\circ(a_i)$, $i = 1, 2, \ldots, n$, $y = \langle b_1, \ldots, b_n \rangle$. Then

$$\|x - y\| = ((a_1 - b_1)^2 + \cdots + (a_n - b_n)^2)^{1/2} \approx 0,$$

that is $x \approx y$. ■

We can conclude from Theorem 3–5.6 that every bounded closed set is compact. In particular, this gives the classical Heine–Borel and Bolzano–Weierstrass theorems (cf. Theorem 3–5.12) for R^n.

2. COMPACT OPERATORS

DEFINITION. T is a *compact operator* if

(1) T is a linear operator on the normed linear space $\mathcal{N}$,
(2) T is a compact map on $\mathcal{N}$.

We immediately have, from Theorem 3–8.1,

THEOREM 2.1. The linear operator T is compact if and only if T maps finite points of $^*\mathcal{N}$ into near-standard points.

In particular, since near-standard points are finite, we have by Theorem 1.8,

COROLLARY 2.2. If the linear operator T is compact, then T is bounded.

THEOREM 2.3. Let T_n be a sequence of compact operators mapping the *Banach space* $\mathcal{N}$ into itself. Let $\|T_n - T\| \to 0$. Then T is compact.

PROOF. Let $B = \{x \mid \|x\| \leq 1\}$. By Theorem 1.12, $T_n \rightrightarrows T$ on B, and hence by Theorem 3–8.2, T is a compact map of B into $\mathcal{N}$. We use this to show that T is a compact operator on $\mathcal{N}$.

Let x be a finite point in $^*\mathcal{N}$. We need only show that Tx is near standard. Let $r \in R^+$ be such that $\|x\| \leq r$. Let $y = x/r$. Then $\|y\| = \|x\|/r \leq 1$, that is, $y \in {}^*B$. Since T is compact on B, Ty is near standard, say, $Ty \approx z$ for $z \in \mathcal{N}$. Then

$$Tx = rTy \approx rz,$$

so that Tx is indeed near standard. ■

We conclude this section by considering an important family of compact operators—the Fredholm integral operators. These operators are usually shown to be compact by using the Arzela–Ascoli Theorem. Instead, we use the same Approximation Theorem (3–7.6) that was used in our proof of the Arzela–Ascoli Theorem.

Let $\mathscr{C}$ be the set of real-valued continuous functions on $[0,1]$. For $f \in \mathscr{C}$, we define $\|f\| = \max_{0 \leq x \leq 1} |f(x)|$. It is very well known and very easy to see that $\mathscr{C}$ forms a Banach space under the natural meaning of $f + g$ and αf, and with this norm. (In particular, to verify completeness, if $\{f_n | n \in N^+\}$ is a Cauchy sequence so that $\|f_n - f_m\| \to 0$ as $m,n \to \infty$, then since we have $|f_n(x) - f_m(x)| \leq \|f_n - f_m\|$ for all $x \in [0,1]$, we have that $f_n \rightrightarrows f$ for some $f \in \mathscr{C}$.)

Let K be some fixed continuous map of $[0,1] \times [0,1]$ into R. The *Fredholm Operator with kernel* K is the mapping T that associates with each $\phi \in \mathscr{C}$, the function $T\phi = \psi$ such that

$$\psi(x) = \int_0^1 K(x,t)\phi(t)\,dt.$$

We shall prove that T is a compact linear operator that maps $\mathscr{C}$ into $\mathscr{C}$. We first show that if $\psi = T\phi$, then $\psi \in \mathscr{C}$. In fact, for any $x,y \in [0,1]$,

$$\begin{aligned} |\psi(x) - \psi(y)| &= \left| \int_0^1 [K(x,t) - K(y,t)]\phi(t)\,dt \right| \\ &\leq \|\phi\| \cdot \int_0^1 |K(x,t) - K(y,t)|\,dt \\ &\leq \|\phi\| \cdot \max_{0 \leq t \leq 1} |K(x,t) - K(y,t)|. \end{aligned}$$

Hence if $x,y \in {}^*[0,1]$, and $x \approx y$, we have $\psi(x) \approx \psi(y)$. That is, ψ is microcontinuous on ${}^*[0,1]$, and, therefore, $\psi \in \mathscr{C}$. This shows that T maps $\mathscr{C}$ into $\mathscr{C}$.

Next we examine ${}^*\mathscr{C}$. We have that $\mathscr{C} = \{f \in U | \vDash \alpha(\mathbf{f})\}$ where $\alpha(\mathbf{v})$ is the formula:

v maps [0,1] into R and v is continuous on [0,1].

Thus ${}^*\mathscr{C} = \{f \in {}^*U | {}^* \vDash {}^*\alpha(\mathbf{f})\}$. That is, f is in ${}^*\mathscr{C}$ just in case f is an internal map of ${}^*[0,1]$ into ${}^*R$, which is everywhere $*$-continuous. Then, ${}^*T$ maps ${}^*\mathscr{C}$ into ${}^*\mathscr{C}$. By the Transfer Principle, the inequality of the previous paragraph yields

$$|\psi(x) - \psi(y)| \leq \|\phi\| \cdot \max_{t \in {}^*[0,1]} |K(x,t) - K(y,t)|$$

for $\phi \in {}^*\mathscr{C}$, $\psi = T\phi$. Hence we have

LEMMA 1. Let ϕ in ${}^*\mathscr{C}$ be finite (i.e., $\|\phi\|$ is finite). Let $\psi = T\phi$. Then ψ is microcontinuous on ${}^*[0,1]$.

T is clearly a linear transformation of $\mathscr{C}$ into itself, since

$$T(\phi_1 + \phi_2) = \int_0^1 K(x,t)(\phi_1(t) + \phi(t_2))\,dt$$
$$= \int_0^1 K(x,t)\phi_1(t)\,dt + \int_0^1 K(x,t)\phi_2(t)\,dt$$
$$= T\phi_1 + T\phi_2.$$
$$T(\alpha\phi) = \int_0^1 K(x,t)\alpha\phi(t)\,dt$$
$$= \alpha \int_0^1 K(x,t)\phi(t)\,dt$$
$$= \alpha T\phi.$$

Also, if $\psi = T\phi$, then for all $x \in [0,1]$,

$$|\psi(x)| \leq \|\phi\| \int_0^1 |K(x,t)|\,dt$$
$$\leq \|\phi\| \cdot \max_{0 \leq t \leq 1} |K(x,t)|$$
$$\leq M \cdot \|\phi\|,$$

where

$$M = \max_{\substack{0 \leq t \leq 1 \\ 0 \leq x \leq 1}} |K(x,t)|.$$

Thus $\|\psi\| \leq M \cdot \|\phi\|$, so that T is bounded and $\|T\| \leq M$.

Finally, we prove

THEOREM 2.4. T is a compact linear operator on $\mathscr{C}$.

PROOF. By Theorem 2.1, we need only show that if $\phi \in {}^*\mathscr{C}$ is finite, then $T\phi$ is near standard. Let $\psi = T\phi$. By Lemma 1, ψ is microcontinuous. Using this we have

LEMMA 2. $\psi(x)$ is near standard for each $x \in {}^*[0,1]$.

PROOF. Using the Transfer Principle,

$$|\psi(x)| \leq \|\psi\| \leq \|T\| \cdot \|\phi\|.$$

Hence $\psi(x)$ is finite and so is near standard for each $x \in {}^*[0,1]$. ■

Using Lemma 2, we can apply the Approximation Theorem (3–7.6) to complete the proof of the theorem. We set $\tilde{\psi}(x) = {}^\circ(\psi(x))$ for $x \in [0,1]$ and conclude that $\tilde{\psi}(x) \approx \psi(x)$ for all $x \in {}^*[0,1]$. Thus

$$\|\tilde{\psi} - \psi\| = \max_{x \in {}^*[0,1]} |\tilde{\psi}(x) - \psi(x)| \approx 0.$$

Hence $\psi \approx \tilde{\psi}$ and so is near standard. ■

3. INTEGRATION OF BANACH SPACE VALUED FUNCTIONS

Let $J = [a,b]$ where $a,b \in R$, $a < b$, and let $\mathscr{B}$ be a Banach space. We write $J^- = J - \{b\}$. Then we write $\mathscr{C}[\mathscr{B}]$ for the set of continuous functions from J into $\mathscr{B}$. As is easily seen, $\mathscr{C}[\mathscr{B}]$ is itself a Banach space with $f + g$ and f having their natural meanings and with $\|f\| = \sup_{x \in J} \|f(x)\| = \max_{x \in J} \|f(x)\|$. We develop the theory of the integral $\int_a^b$ as a linear operator from $\mathscr{C}[\mathscr{B}]$ into $\mathscr{B}$.

DEFINITION. A step function on J is a map s of J^- into $\mathscr{B}$ such that

$$s(x) = r_{i-1} \quad \text{for} \quad x_{i-1} \leq x < x_i,\ i = 1, \ldots, n,$$

where $a = x_0 < x_1 < \cdots < x_n = b$ and $\{r_i | 0 \leq i < n\}$ is a finite sequence of points of $\mathscr{B}$.

We begin with the integral of a step function.

DEFINITION. Let s be a step function on J. Then

$$\int_a^b s(x)\,dx = \sum_{i=1}^{n} r_{i-1}(x_i - x_{i-1}).$$

It is important to note that $\int_a^b s(x)\,dx$ depends only on s and not on the particular partition used to define s. Thus let

$$s(x) = r_{i-1} \quad \text{for} \quad x_{i-1} \leq x < x_{i+1},\ i = 1, \ldots, n,$$

where $a = x_0 < x_1 < \cdots < x_n = b$. Then the same function s will be defined if the point ξ is inserted between x_{j-1} and x_j for some fixed j so that

$$s(x) = \begin{cases} r_{j-1} & \text{for} \quad x_{j-1} \leq x < \xi \\ r_{j-1} & \text{for} \quad \xi < x \leq x^j. \end{cases}$$

Now the term $r_{j-1}(x_j - x_{j-1})$ is then replaced by $r_{j-1}(\xi - x_{j-1}) + r_{j-1}(x_j - \xi)$ in the sum defining $\int_a^b s(x)\,dx$, but this does not change the sum. Iterating this process, we see that adding any finite number of points to the partition leaves the value of $\int_a^b s(x)\,dx$ unchanged. Finally, if s is given by two different partitions, the partition obtained by combining the points of both must yield the same value for the integral as obtained for each one.

Let $\mathscr{S}$ be the set of all step functions on J. For any $s \in \mathscr{S}$, we define the *norm* of s as

$$\|s\| = \sup_{x \in J^-} \|s(x)\| = \max_{0 \leq i \leq n} \|r_i\|.$$

LEMMA 1. $\mathscr{S}$ is a normed linear space under the natural meanings of $s + t$ and αs and with norm defined as above. Also, $\int_a^b$ is a bounded linear operator from $\mathscr{S}$ into $\mathscr{B}$.

PROOF. Let s and t be step functions. Then $s + t$ and αs are obviously step functions.

To check the norm properties we have

$$\begin{aligned}\|s + t\| &= \sup_{x \in J} \|s(x) + t(x)\| \\ &\le \sup_{x \in J} \{\|s(x)\| + \|t(x)\|\} \\ &\le \sup_{x \in J} \|s(x)\| + \sup_{x \in J} \|t(x)\| \\ &= \|s\| + \|t\|. \\ \|\alpha s\| &= \sup_{x \in J} \|\alpha s(x)\| \\ &= |\alpha| \sup_{x \in J} \|s(x)\| = |\alpha| \cdot \|s\|.\end{aligned}$$

Also, $\|s\| = 0$ obviously implies $s(x) = 0$ for all x, that is, $s = 0$.

Clearly,

$$\int_a^b \alpha s(x)\,dx = \sum_{i=1}^{n} \alpha r_{i-1}(x_i - x_{i-1}) = \alpha \int_a^b s(x)\,dx.$$

That $\int_a^b (s(x) + t(x))\,dx = \int_a^b s(x)\,dx + \int_a^b t(x)\,dx$ follows similarly if we use the same partition in defining both functions. Finally,

$$\begin{aligned}\left\|\int_a^b s(x)\,dx\right\| &= \left\|\sum_{i=1}^{n} r_{i-1}(x_i - x_{i-1})\right\| \\ &\le \sum_{i=1}^{n} \|r_{i-1}\|(x_i - x_{i-1}) \\ &\le \|s\| \sum_{i=1}^{n} (x_i - x_{i-1}) \\ &= (b - a)\|s\|.\end{aligned}$$

Hence $\int_a^b$ is a bounded linear operator on $\mathscr{S}$. ∎

Next, we need to calculate $^*\mathscr{S}$. To do so we need to be very precise about $\mathscr{S}$. For each $n \in N$, let

$$\sigma_n = \{m \in N | m \le n\}.$$

Thus σ maps N into $\mathscr{P}(N)$, so that $\sigma \in U$. A finite sequence of elements of a set A (where $A \subseteq S$) is just a map of σ_n into A for some $n \in N$. Now for each $s \in \mathscr{S}$, there is an $n \in N^+$ and a pair of finite sequences: x, which maps σ_n into J such that $x_0 = a$, $x_n = b$, and $x_i < x_{i+1}$ for $0 \leq i < n$; and r, which maps σ_{n-1} into $\mathscr{B}$. Finally, $s(x) = r_{i-1}$ for $x_{i-1} \leq x < x_i$, $1 \leq i \leq n$.

Now $^*\sigma$ maps $^*N$ into $^*\mathscr{P}(N)$, and using Theorem 1–7.12 and the Transfer Principle, it follows that $(^*\sigma)_n = \{m \in {}^*N | m \leq n\}$ for all $n \in {}^*N$. Since $(^*\sigma)_n = \sigma_n$ for $n \in N$, we can, as usual, omit the *. (What is different in this case, is that $\sigma[N] \not\subseteq S$.) Using the Transfer Principle, we conclude that each element s of $^*\mathscr{S}$ is given by

(1) an element $n \in {}^*N^+$;
(2) an *internal* map x of σ_n into $^*J$ such that $x_0 = a$, $x_n = b$, and $x_i < x_{i+1}$ for $0 \leq i < n$;
(3) an *internal* map r of σ_{n-1} into $^*\mathscr{B}$ such that $s(x) = r_{i-1}$ for $x \in {}^*J$, $x_{i-1} \leq x < x_i$, $1 \leq i \leq n$.

For each $s \in {}^*\mathscr{S}$, we have (by the Transfer Principle)

(1) $\|s(x)\| \leq \|s\|$ for each $x \in {}^*J^-$;
(2) for each positive hyperreal number δ, there is an $x \in {}^*J^-$ such that $\|s(x)\| > \|s\| - \delta$.

Thus $\|s\| \approx 0$ if and only if $s(x) \approx 0$ for all $x \in {}^*J^-$. Then, for $s_1, s_2 \in {}^*\mathscr{S}$, $s_1 \approx s_2$ if and only if $\|s_1 - s_2\| \approx 0$, that is, if and only if $s_1(x) \approx s_2(x)$ for $x \in {}^*J^-$.

Since $\int_a^b$ is a map of $\mathscr{S}$ into $\mathscr{B}$, $^*\int_a^b$ maps $^*\mathscr{S}$ into $^*\mathscr{B}$. Let $s \in {}^*\mathscr{S}$ and let $s(x) = r_{i-1}$ for $x_{i-1} \leq x < x_i$, $1 \leq i \leq n$, where $n \in {}^*N$ and where r and x are internal sequences. Then by the Transfer Principle,

$$^*\int_a^b s(x)\, dx = \sum_{i=1}^{n} r_{i-1}(x_i - x_{i-1}).$$

In particular, $^*\int_a^b$ and $\int_a^b$ give the same value on $\mathscr{S}$. Hence we omit the * from $\int_a^b$.

Since $\int_a^b$ is a bounded linear operator, $\int_a^b$ is microcontinuous on $^*\mathscr{S}$. That is,

$$s_1 \approx s_2 \quad \text{implies} \quad \int_a^b s_1(x)\, dx \approx \int_a^b s_2(x)\, dx.$$

The key result is

THEOREM 3.1 Let $f \in \mathscr{C}[\mathscr{B}]$. Then there is an $s \in {}^*\mathscr{S}$ such that $f(x) \approx s(x)$ for all $x \in {}^*J^-$, and $\int_a^b s(x)\, dx$ is near standard.

PROOF. As in the proof of Theorem 2–5.5, we consider the map t of $N \times N$ into J such that

$$t(n,i) = t_i^{(n)} = a + \frac{b - a}{n} i \quad \text{for} \quad 0 \le i \le n.$$

Next, we set $s(n,x) = s_n(x) = f(t_i^{(n)})$ for $t_i^{(n)} \le x < t_{i+1}^{(n)}$, $0 \le i < n$. Then for each $n \in {}^*N^+$, $s_n \in {}^*\mathscr{S}$. In particular, for $\nu \in {}^*N - N$, $t_{i+1}^{(\nu)} - t_i^{(\nu)} = \dfrac{b - a}{\nu} \approx 0$, so that $t_i^{(\nu)} \le x < t_{i+1}^{(\nu)}$ implies $x \approx t_i^{(\nu)}$. Hence (since f is uniformly continuous on the compact set J) $f(x) \approx f(t_i^{\nu}) = s_\nu(x)$. It remains only to show that $\int_a^b s_\nu(x)\,dx$ is near standard.

Let $I_n = \int_a^b s_n(x)\,dx$ for $n \in N^+$. By the Transfer Principle, $I_n = \int_a^b s_n(x)\,dx$ for $n \in {}^*N^+$. Now, let $\nu,\mu \in {}^*N - N$. Then by what we have already proved, $s_\nu(x) \approx f(x) \approx s_\mu(x)$ for $x \in {}^*J^-$. Hence $s_\nu \approx s_\mu$. By the microcontinuity of $\int_a^b$, we have $I_\nu \approx I_\mu$. That is, $\{I_n | n \in N^+\}$ is a Cauchy sequence. Since $\mathscr{B}$ is a Banach space (i.e., it is complete), there is a point $I \in \mathscr{B}$ such that $I_\nu \approx I$ for all $\nu \in {}^*N - N$. Thus I_ν is near standard. ■

THEOREM 3.2. Let $s_1,s_2 \in {}^*\mathscr{S}$, $f \in \mathscr{C}[\mathscr{B}]$, where $s_1(x) \approx f(x)$, $s_2(x) \approx f(x)$ for $x \in {}^*J^-$. Then $\int_a^b s_1(x)\,dx \approx \int_a^b s_2(x)\,dx$.

PROOF. The hypothesis implies that $s_1 \approx s_2$. The conclusion then follows from the microcontinuity of $\int_a^b$. ■

The two previous theorems justify the following:

DEFINITION. Let $f \in \mathscr{C}[\mathscr{B}]$. Then

$$\int_a^b f(x)\,dx = {}^\circ\!\left(\int_a^b s(x)\,dx\right),$$

where s is any element of ${}^*\mathscr{S}$ for which $s(x) \approx f(x)$ for all $x \in {}^*J^-$.

We will need the

LEMMA 2. If $f \in \mathscr{C}[\mathscr{B}]$, $s \in {}^*\mathscr{S}$ and $s(x) \approx f(x)$ for $x \in {}^*J^-$, then $\|f\| = {}^\circ\|s\|$.

PROOF. Let $\|s\| = \|s(\xi)\|$ for a suitable $\xi \in {}^*J^-$. Let $t = {}^\circ\xi$. Then $f(t) \approx f(\xi) \approx s(\xi)$. Hence $\|f(t)\| \approx \|s\|$, that is, $\|f(t)\| = {}^\circ\|s\|$. Now let $x \in J^-$. Then $\|s\| \ge \|s(x)\|$. Taking standard parts: $\|f(t)\| \ge \|f(x)\|$. Finally, by continuity, this last inequality is also correct for $x = b$, that is, for all $x \in J$. Hence $\|f\| = \|f(t)\| = {}^\circ\|s\|$. ■

THEOREM 3.3. $\int_a^b$ is a bounded linear operator on $\mathscr{C}[\mathscr{B}]$. In fact we have, $\|\int_a^b f(x)\,dx\| \le (b - a)\|f\|$.

PROOF. Let $f_1, f_2 \in \mathscr{C}[\mathscr{B}]$. Let $f_1(x) \approx s_1(x)$, $f_2(x) \approx s_2(x)$ for $x \in {}^*J^-$ where $s_1, s_2 \in {}^*\mathscr{C}$. Then $f_1(x) + f_2(x) \approx s_1(x) + s_2(x)$, so that

$$\begin{aligned}\int_a^b [f_1(x) + f_2(x)]\,dx &= {}^\circ\left(\int_a^b [s_1(x) + s_2(x)]\,dx\right)\\ &= {}^\circ\left(\int_a^b s_1(x)\,dx\right) + {}^\circ\left(\int_a^b s_2(x)\,dx\right)\\ &= \int_a^b f_1(x)\,dx + \int_a^b f_2(x)\,dx.\end{aligned}$$

Likewise,

$$\begin{aligned}\int_a^b \alpha f_1(x)\,dx &= {}^\circ\left(\int_a^b \alpha s_1(x)\,dx\right)\\ &= \alpha\,{}^\circ\left(\int_a^b s_1(x)\,dx\right) = \alpha \int_a^b f_1(x)\,dx.\end{aligned}$$

Finally, we have using Lemmas 1 and 2,

$$\begin{aligned}\left\|\int_a^b f_1(x)\,dx\right\| &= {}^\circ\left\|\int_a^b s_1(x)\,dx\right\|\\ &\le (b - a)\cdot {}^\circ\|s_1\|\\ &= (b - a)\|f_1\|. \quad \blacksquare\end{aligned}$$

THEOREM 3.4. $\int_a^b k\,dx = k(b - a)$.

PROOF. For $f(x) = k$, f is a step function. ■

THEOREM 3.5. Let α be a continuous function from J into R, and let $k \in \mathscr{B}$. Then

$$\int_a^b k\alpha(x)\,dx = k \int_a^b \alpha(x)\,dx.$$

PROOF. Let $s(x) = r_{i-1}$ for $x_{i-1} \le x < x_i$, $1 \le i \le \nu$, $s(x) \approx \alpha(x)$ for all $x \in J^-$, so that $ks(x) \approx k\alpha(x)$ as well, and

$$\begin{aligned}\int_a^b k\alpha(x)\,dx &\approx \int_a^b ks(x)\,dx = \sum_{i=1}^{\nu} kr_{i-1}(x_i - x_{i-1})\\ &= k\sum_{k=1}^{\nu} r_{i-1}(x_i - x_{i-1}) \approx k\int_a^b \alpha(x)\,dx. \quad \blacksquare\end{aligned}$$

THEOREM 3.6. Let $u \in \mathscr{C}[\mathscr{B}]$, and let f be a bounded linear functional on $\mathscr{B}$. Then

$$f\left(\int_a^b u(x)\,dx\right) = \int_a^b f(u(x))\,dx.$$

PROOF. Let $u(x) \approx s(x)$ for $x \in {}^*J^-$ where $s \in {}^*\mathscr{S}$, say $s(x) = r_{i-1}$ for $x_{i-1} \leq x < x_i$, $1 \leq i \leq \nu$. Then since f is microcontinuous, $f(u(x)) \approx f(s(x))$ for $x \in {}^*J^-$, and

$$\int_a^b u(x)\,dx \approx \int_a^b s(x)\,dx = \sum_{i=1}^{\nu} r_{i-1}(x_i - x_{i-1}).$$

Since f is continuous and linear,

$$\begin{aligned} f\left(\int_a^b u(x)\,dx\right) &\approx f\left(\sum_{i=1}^{\nu} s_i(x_i - x_{i-1})\right) \\ &= \sum_{i=1}^{\nu} f(s_i)(x_i - x_{i-1}) \\ &= \int_a^b f(s(x))\,dx \\ &\approx \int_a^b f(u(x))\,dx. \quad \blacksquare \end{aligned}$$

The interchangeability of the order of integration for continuous real functions is an immediate consequence of this theorem:

COROLLARY 3.7. Let f be a continuous map of $[a,b] \times [c,d]$ into R. Then

$$\int_c^d \int_a^b f(x,y)\,dx\,dy = \int_a^b \int_c^d f(x,y)\,dy\,dx.$$

PROOF. Let $\mathscr{B} = R$ so $\mathscr{C}[\mathscr{B}]$ is the space of continuous real valued functions on J. Let F be the bounded linear functional in $\mathscr{C}[\mathscr{B}]$ defined by

$$F(\alpha) = \int_c^d \alpha(y)\,dy.$$

The theorem yields

$$F\left(\int_a^b f(x,y)\,dx\right) = \int_a^b F(f(x,y))\,dx.$$

But this is the desired result. $\blacksquare$

4. DIFFERENTIAL CALCULUS

In this section, we generalize the treatment of differential calculus in Chapter 2.

Thus let $\mathscr{N},\mathscr{M}$ be normed linear spaces over the same field D. (As usual, D is R or C.) Let $I_{\mathscr{N}},I_{\mathscr{M}}$ be the set of infinitesimals of $\mathscr{N}$ and of $\mathscr{M}$, respectively.

We note the obvious closure properties:

(1) $x,y \in I_{\mathscr{N}}$ implies $x + y \in I_{\mathscr{N}}$;
(2) if $x \in I_{\mathscr{N}}$ and $\alpha \in {}^*D$ is finite, then $\alpha x \in I_{\mathscr{N}}$.

DEFINITION. Let f be an internal map whose domain and range are subsets of ${}^*\mathscr{N}$ and ${}^*\mathscr{M}$, respectively. Then f is called a *local map* if $f[I_{\mathscr{N}}] \subseteq I_{\mathscr{M}}$.

If f and g are local maps, we write $f \sim g$ and say that f and g are *equivalent* if for all $x \in I_{\mathscr{N}}$ for which $x \neq 0$, we have

$$\frac{f(x) - g(x)}{\|x\|} \approx 0.$$

It is obvious that $\sim$ is an equivalence relation.

As in the special case $\mathscr{N} = \mathscr{M} = R$, this condition can be written

$$f(x) = g(x) + \alpha(x) \cdot \|x\|,$$

where $\alpha \approx 0$ for all $x \in I_{\mathscr{N}}$, that is, where α is itself a local map.

DEFINITION. The local map f is called *locally linear* if

(1) $x,y \in I_{\mathscr{N}}$ implies $f(x + y) = f(x) + f(y)$,
(2) if $x \in I_{\mathscr{N}}$ and $\alpha \in {}^*D$ is finite, then $f(\alpha x) = \alpha f(x)$.

THEOREM 4.1. Let f be locally linear. Then there is a unique internal linear operator T of ${}^*\mathscr{N}$ into ${}^*\mathscr{M}$ such that $f(x) = Tx$ for $x \in I_{\mathscr{N}}$. Moreover, $\|T\|$ is finite.

PROOF. Let α,β be infinitesimal nonzero elements of ${}^*D$. Let $x \in {}^*\mathscr{N}$ such that $\alpha x,\beta x \approx O$. Then

$$\frac{f(\alpha x)}{\alpha} = \frac{\beta f(\alpha x)}{\alpha\beta} = \frac{f(\alpha\beta x)}{\alpha\beta} = \frac{\alpha f(\beta x)}{\alpha\beta} = \frac{f(\beta x)}{\beta}. \qquad (\ddagger)$$

Let ν be some fixed infinite integer. Then, $x/\nu\|x\| \approx 0$ for all nonzero $x \in {}^*\mathscr{N}$. We define $Tx = \nu\|x\|f(x/\nu\|x\|)$ for $x \neq 0$; $T0 = 0$. Thus T is internal (by Theorem 1–8.9). Moreover, ($\ddagger$) shows that $Tx = f(\alpha x)/\alpha$ for each infinitesimal α such that $\alpha x \approx 0$.

In particular, if $x \in I_{\mathscr{N}}$, we have, for any $\alpha \approx 0$,

$$Tx = \frac{f(\alpha x)}{\alpha} = \frac{\alpha f(x)}{\alpha} = f(x).$$

We next verify that T is linear: Let $x,y \in {}^*\mathcal{N}$; let $\alpha \in {}^*D$ be chosen so that $\alpha \approx 0$, $\alpha x, \alpha y \approx 0$. Then

$$\begin{aligned} T(x + y) &= \frac{1}{\alpha}\{f(\alpha x + \alpha y)\} \\ &= \frac{1}{\alpha}\{f(\alpha x) + f(\alpha y)\} \\ &= Tx + Ty. \end{aligned}$$

Also, for any $\lambda \in {}^*D$, choose $\alpha \approx 0$ so that $\alpha x, \alpha\lambda x \approx 0$. Then

$$T(\lambda x) = \frac{1}{\alpha} f(\alpha\lambda x) = \frac{\lambda}{\alpha} f(\alpha x) = \lambda Tx.$$

To verify uniqueness, let $f(x) = T_1 x = T_2 x$ for $x \in I_{\mathcal{N}}$ where T_1, T_2 are linear operators on ${}^*\mathcal{N}$. Let x be any point in ${}^*\mathcal{N}$. Then

$$T_1 x = v\|x\| T_1\left(\frac{x}{v\|x\|}\right) = v\|x\| T_2\left(\frac{x}{v\|x\|}\right) = T_2 x.$$

Finally, suppose $\|T\|$ is infinite. Then (by the Transfer Principle) for some $x \in {}^*\mathcal{N}$ with $\|x\| = 1$, $\|Tx\| \geq \|T\| - 1$, that is, $\|Tx\|$ must be infinite. Thus $x/\|Tx\| \approx 0$, and we have

$$Tx = \|Tx\| f\left(\frac{x}{\|Tx\|}\right),$$

that is,

$$\|Tx\| = \|Tx\| \cdot \left\| f\left(\frac{x}{\|Tx\|}\right)\right\|.$$

Then $1 = \|f(x/\|Tx\|)\| \notin I_{\mathcal{N}}$, a contradiction. ■

THEOREM 4.2. Let f_1, f_2 be locally linear maps, and let T_1, T_2 be related to f_1, f_2, respectively, as in the previous theorem. Then $f_1 \sim f_2$ if and only if $T_1 x \approx T_2 x$ for all $x \in {}^*\mathcal{N}$ with $\|x\| = 1$.

PROOF. First let $T_1 x \approx T_2 x$ for all $x \in {}^*\mathcal{N}$ with $\|x\| = 1$. Let $u \in I_{\mathcal{N}}$. Then

$$\begin{aligned} \frac{f_1(u) - f_2(u)}{\|u\|} &= \frac{T_1(u) - T_2(u)}{\|u\|} \\ &= T_1\left(\frac{u}{\|u\|}\right) - T_2\left(\frac{u}{\|u\|}\right) \\ &\approx 0. \end{aligned}$$

Hence $f_1 \sim f_2$.

Conversely, if $f_1 \sim f_2$ and $\|x\| = 1$, then choosing $\alpha \approx 0$, $\alpha > 0$, we have

$$T_1x - T_2x = \frac{1}{\alpha}[T_1(\alpha x) - T_2(\alpha x)]$$

$$= \frac{f_1(\alpha x) - f_2(\alpha x)}{\|\alpha x\|} \approx 0. \quad \blacksquare$$

DEFINITION. A locally linear map f is called a *differential* if there is a bounded linear transformation T mapping $\mathcal{N}$ into $\mathcal{M}$ such that $Tx = f(x)$ for $x \in I_{\mathcal{N}}$.

THEOREM 4.3. If f_1, f_2 are differentials and $f_1 \sim f_2$, then $f_1 = f_2$.

PROOF. We have standard linear operators T_1, T_2 such that $f_1x = T_1x$, $f_2x = T_2x$, for $x \approx 0$. By Theorem 4.2, $T_1u \approx T_2u$ for $\|u\| = 1$. For u a standard point of $\mathcal{N}$ with $\|u\| = 1$, this becomes $T_1u = T_2u$. Hence for any $x \in \mathcal{N}$, if $x \neq 0$,

$$T_1x = \|x\| \cdot T_1\left(\frac{x}{\|x\|}\right) = \|x\| \cdot T_2\left(\frac{x}{\|x\|}\right) = T_2x.$$

Finally, since $T_1 = T_2$, we have $f_1 = f_2$. $\blacksquare$

DEFINITION. Let f be any map of $\mathcal{N}$ into $\mathcal{M}$, and let $x_0 \in \mathcal{N}$. Then we write $\Delta_{x_0}f$ for the map on $I_{\mathcal{N}}$ defined by

$$\Delta_{x_0}f(u) = f(x_0 + u) - f(x_0).$$

We have at once

COROLLARY 4.4. f is continuous at x_0 if and only if $\Delta_{x_0}f$ is a local map.

DEFINITION. Let f map $\mathcal{N}$ into $\mathcal{M}$, and let $x_0 \in \mathcal{N}$. Then f is *differentiable at* x_0 if $\Delta_{x_0}f$ is equivalent to a differential.

In this case, by Theorem 4.3, there is a *unique* differential to which $\Delta_{x_0}f$ is equivalent. We write this $d_{x_0}f$ or just df, and note that

$$\Delta_{x_0}f(u) = d_{x_0}f(u) + \alpha(u) \cdot \|u\|,$$

where $\alpha(u) \approx 0$ for $u \approx 0$, (i.e., α is a local map).

We now consider some classical special cases. First, let $\mathcal{N} = R^n$, $n > 0$ and $\mathcal{M} = R$. Then a map f of $\mathcal{N}$ into $\mathcal{M}$ is simply a real-valued function of n real variables; a linear operator of $\mathcal{N}$ into $\mathcal{M}$ is given by a form: $r_1x_1 + r_2x_2 + \cdots + r_nx_n$. To say that f is differentiable at a point

$a = \langle a_1 \ldots, a_n \rangle$ is then to say that

$$f(a_1 + h_1, \ldots, a_n + h_n) - f(a_1, \ldots, a_n) = r_1 h_1 + r_2 h_2 + \cdots + r_n h_n + \|h\| \cdot \alpha(h),$$

where $h = \langle h_1, \ldots, h_n \rangle$ and $\alpha(h) \approx 0$ for $h \approx 0$. The numbers $r_1, \ldots, r_n$ are then the respective *partial derivatives* of f with respect to $x_1, \ldots, x_n$ at the point a. Details are left to the reader.

If $\mathcal{N} = R^n$, $\mathcal{M} = R^m$, $n,m > 0$, and f is a differentiable map of $\mathcal{N}$ into $\mathcal{M}$ at the point $\alpha \in \mathcal{N}$, then the differential of f is associated with an n by m matrix, the Jacobian matrix of the corresponding linear operator. Details are again left to the reader.

5
HILBERT SPACE

1. UNITARY SPACES

In this chapter, we apply nonstandard methods to obtain several important theorems about Hilbert space. One of these (the Bernstein–Robinson theorem on the existence of invariant subspaces) was proved for the first time using nonstandard methods, settling an important problem that had been open for a long time.

We freely cite results from finite dimensional linear algebra, using Halmos [5] as a reference. However, we assume no previous familiarity with the infinite dimensional theory.

As usual, we write $\bar{z}$ for the conjugate of $z \in C$, where C as above is the field of complex numbers.

DEFINITION. A *unitary space* H is a linear space over C on which is defined an *inner product* (a,b) that is a map of $H \times H$ into C that satisfies:

(1) $(a,a) \geq 0$ (so $(a,a) \in R$); $(a,a) = 0$ if and only if $a = 0$;
(2) $(a,b) = \overline{(b,a)}$;
(3) $(a + b,c) = (a,c) + (b,c)$;
(4) $(\lambda a,b) = \lambda(a,b)$ for $\lambda \in C$.

It follows at once from (2), (3), and (4) that we have also

(5) $(a,b + c) = (a,b) + (a,c)$,
(6) $(a,\lambda b) = \bar{\lambda}(a,b)$.

We write $\|x\| = (x,x)^{1/2}$ for any $x \in H$; it is then not difficult to verify (e.g., see Halmos [5]) that $\|\cdots\|$ is a norm (so a unitary space is a normed linear space) and that the Schwarz inequality

$$|(x,y)| \leq \|x\| \cdot \|y\|$$

holds.

The Schwarz inequality implies that the inner product is a continuous map of $H \times H$ into C since

$$\begin{aligned}\|(x,y) - (u,v)\| &= \|(x,y) - (x,v) + (x,v) - (u,v)\| \\ &\leq \|(x,y - v)\| + \|(x - u,v)\| \\ &\leq \|x\| \cdot \|y - v\| + \|x - u\| \cdot \|v\|.\end{aligned}$$

If a,b are points of a unitary space, we write $a \perp b$ and say that a is *perpendicular* to b to mean that $(a,b) = 0$. A sequence $\{e_n | n \in N^+\}$ of points is called *orthonormal* if

(a) $$\|e_n\| = 1,$$

(b) $$e_n \perp e_m \quad \text{for} \quad n \neq m.$$

This condition can be expressed by the single equation:

$$(e_n,e_m) = \delta^n_m,$$

where δ^n_m ("Kronecker's delta") $= 0$ if $n \neq m$ and $=1$ if $n = m$.

An orthonormal sequence is obviously linearly independent. For,

$$\sum_{i=1}^{n} \alpha_i e_i = 0$$

implies that for $1 \leq j \leq n$,

$$\alpha_j = \sum_{i=1}^{n} \alpha_i (e_i,e_j) = (0,e_j) = 0.$$

A linear operator T on a unitary space is called *Hermitian* or *self-adjoint* if for all x,y in the space

$$(Tx,y) = (x,Ty).$$

In particular, if T is Hermitian,

$$(Tx,x) = (x,Tx) = \overline{(Tx,x)},$$

so that (Tx,x) is a *real number* for each x in the space.

Let T_1,T_2 be Hermitian operators on the same space. Then we write $T_1 \leq T_2$ to mean that for all x in the space,

$$(T_1x,x) \leq (T_2x,x).$$

As a first example of a unitary space, consider the n-dimensional linear space C^n whose points are n-tuples of complex numbers, added and multiplied by complex numbers component-wise. On this space we can define the inner product of $a = \langle \alpha_1, \ldots, \alpha_n \rangle$, $b = \langle \beta_1, \ldots, \beta_n \rangle$ by

$$(a,b) = \sum_{i=1}^{n} \alpha_i \overline{\beta_i}$$

and easily verify that axioms (1) to (4) are satisfied. Since C^n as thus defined is a unitary space, the Schwarz inequality yields

$$\left|\sum_{i=1}^{n} \alpha_i \overline{\beta_i}\right| \leq \left(\sum_{i=1}^{n} |\alpha_i|^2 \cdot \sum_{i=1}^{n} |\beta_i|^2\right)^{1/2}.$$

Our next example (which will be important in this chapter) is the space l^2, which is an infinite dimensional analog of C^n. The elements of l^2 are all the infinite sequences:

$$a = (\alpha_1,\alpha_2, \ldots),$$

such that

$$\sum_{i=1}^{\infty} |\alpha_i|^2 < \infty.$$

If a and $b = (\beta_1,\beta_2, \ldots)$ are points of l^2 and $\lambda \in C$, we define

$$a + b = (\alpha_1 + \beta_1, \alpha_2 + \beta_2, \ldots),$$
$$\lambda a = (\lambda\alpha_1,\lambda\alpha_2, \ldots).$$

That $a + b \in l^2$ is easy to see since by the triangle inequality for C^n (also known as Minkowski's inequality),

$$\left(\sum_{i=1}^{n} |\alpha_i + \beta_i|^2\right)^{1/2} \leq \left(\sum_{i=1}^{n} |\alpha_i|^2\right)^{1/2} + \left(\sum_{i=1}^{n} |\beta_i|^2\right)^{1/2}.$$

We also wish to define $(a,b) = \sum_{i=1}^{\infty} \alpha_i \overline{\beta}_i$. To see that this is legitimate note that, applying Schwarz' inequality for C^n to the points $\langle|\alpha_1|, \ldots ,|\alpha_n|\rangle$, $\langle|\beta_1|, \ldots ,|\beta_n|\rangle$, we have

$$\sum_{i=1}^{n} |\alpha_i \beta_i| \leq \left(\sum_{i=1}^{n} |\alpha_i|^2 \cdot \sum_{i=1}^{n} |\beta_i|^2\right)^{1/2},$$

so that the series $\sum_{i=1}^{\infty} \alpha_i \overline{\beta}_i$ is absolutely convergent. It is easily verified that under these definitions l^2 forms a unitary space.

Consider the sequence of points of l^2, $\{b_k | k \in N^+\}$, where for each k, $b_k = \{\delta_n^k | n \in N^+\}$. Thus

$$b_k = (0,0, \ldots ,0,1,0, \ldots),$$

where 1 occurs as the kth component. The sequence $\{b_k\}$ clearly is orthonormal and hence is linearly independent. Let $a = (\alpha_1,\alpha_2, \ldots) \in l^2$. Then

$$\left\|a - \sum_{i=1}^{n} \alpha_i b_i\right\|^2 = \sum_{i=n+1}^{\infty} |\alpha_i|^2 = a^2 - \sum_{i=1}^{n} |\alpha_i|^2 \to 0,$$

as $n \to \infty$. Thus

$$a = \sum_{i=1}^{\infty} \alpha_i b_i,$$

and the sequence $\{b_k\}$ is an *orthonormal basis* in l^2. Since $\{b_k\}$ is linearly independent, it follows that l^2 is infinite dimensional.

DEFINITION. A unitary space is called a *Hilbert space* if it is infinite dimensional and (Cauchy) complete.

THEOREM 1.1. l^2 is a Hilbert space.

PROOF. We need to verify completeness. Thus let $\{a_n | n \in N^+\}$ be a Cauchy sequence of points of l^2, say $a_n = \{\alpha_m^{(n)}\}$ for $n = 1,2,3,\ldots$. Let ε be some definite positive real number. Then there is an $n_0 \in N^+$ such that

$$p,q > n_0 \quad \text{implies} \quad |a_p - a_q| < \varepsilon.$$

That is, for $p,q > n_0$,

$$A_{p,q} = \sum_{m=1}^{\infty} |\alpha_m^{(p)} - \alpha_m^{(q)}|^2 < \varepsilon,$$

so that for such p,q and all $K \in N^+$,

$$|\alpha_m^{(p)} - \alpha_m^{(q)}|^2 \leq \sum_{m=1}^{K} |\alpha_m^{(p)} - \alpha_m^{(q)}|^2 \leq A_{p,q} < \varepsilon.$$

Thus for each fixed m, the sequence of complex numbers $\{\alpha_m^{(n)} | n \in N^+\}$ is a Cauchy sequence and hence converges; that is, $\alpha_m^{(n)} \to \alpha_m$ as $n \to \infty$. Hence letting $q \to \infty$, for each $K \in N^+$ and for $p > n_0$,

$$\sum_{m=1}^{K} |\alpha_m^{(p)} - \alpha_m|^2 \leq \varepsilon,$$

which in turn implies (letting $K \to \infty$),

$$\sum_{m=1}^{\infty} |\alpha_m^{(p)} - \alpha_m|^2 \leq \varepsilon.$$

From the convergence of this last series, we conclude that with $p = n_0 + 1$,

$$b = \{\alpha_1^{(p)} - \alpha_1, \alpha_2^{(p)} - \alpha_2, \ldots\} \in l^2.$$

Thus

$$a = \{\alpha_m\} = \{\alpha_m^{(p)}\} - b \in l^2.$$

Finally, the last inequality simply states that

$$p > n_0 \quad \text{implies} \quad \|a_p - a\|^2 \leq \varepsilon.$$

Thus $a_p \to a$, and the space is complete. ■

We have established the existence of an orthonormal basis (namely, $\{b_k\}$) for l^2. Now, and for the rest of this chapter, we let $\{e_k\}$ be any fixed orthonormal basis for l^2. We want to show that the inner product is calculated in exactly the same manner with respect to $\{e_k\}$ as with respect to $\{b_k\}$. Namely,

THEOREM 1.2. Let $a = \sum_{i=1}^{\infty} \alpha_i e_i$, $b = \sum_{i=1}^{\infty} \beta_i e_i$. Then

$$(a,b) = \sum_{i=1}^{\infty} \alpha_i \overline{\beta}_i.$$

PROOF. Using (2) to (6) at the beginning of this section and the fact that $\{e_k\}$ is orthonormal,

$$\begin{aligned} \left(\sum_{i=1}^{n} \alpha_i e_i, \sum_{i=1}^{n} \beta_i e_i\right) &= \sum_{i=1}^{n} \sum_{j=1}^{n} (\alpha_i e_i, \beta_j e_j) \\ &= \sum_{i=1}^{n} \sum_{j=1}^{n} \alpha_i \overline{\beta}_j (e_i, e_j) \\ &= \sum_{i=1}^{n} \alpha_i \overline{\beta}_i. \end{aligned}$$

The result follows from the continuity of the inner product. ■

COROLLARY 1.3. Let $a = \sum_{i=1}^{\infty} \alpha_i e_i$. Then

$$\|a\|^2 = \sum_{i=1}^{\infty} |\alpha_i|^2.$$

DEFINITION. A Hilbert space is called *separable* if it has a countable dense subset.

THEOREM 1.4. l^2 is separable.

PROOF. Let A be the set of all finite linear combinations of the basis vectors e_i with coefficients whose real and imaginary parts are rational. Then A is obviously countable; we show that it is dense in l^2.

Let $b = \sum_{i=1}^{\infty} \alpha_i e_i \in l^2$, and let $\varepsilon > 0$ be real. We then find an $a \in A$ such that $\|b - a\| < \varepsilon$. First choose n so that $\sum_{i=n+1}^{\infty} |\alpha_i|^2 < \frac{2}{3}\varepsilon^2$. Next for each $i = 1,2,\ldots,n$ choose γ_i with rational real and imaginary parts such that $|\alpha_i - \gamma_i| < \varepsilon/2^i$, and set $a = \sum_{i=1}^{n} \gamma_i e_i$. Then,

$$\begin{aligned}\|b - a\|^2 &= \sum_{i=1}^{n} |\alpha_i - \gamma_i|^2 + \sum_{i=n+1}^{\infty} |\alpha_i|^2 \\ &< \varepsilon^2 \left\{ \sum_{i=1}^{n} \frac{1}{4^i} + \frac{2}{3} \right\} \\ &< \varepsilon^2 \left(\frac{1}{3} + \frac{2}{3} \right) = \varepsilon^2. \quad \blacksquare\end{aligned}$$

Now the famous Riesz–Fischer theorem states that any two separable Hilbert spaces are isomorphic (with the isomorphism preserving the inner product). Hence what we prove about l^2 holds for any separable Hilbert space. In fact, we do not explicitly use the Riesz–Fischer theorem and, therefore, when appropriate, limit our statements to l^2.

It is useful to assume that $C \subseteq S$ and $l^2 \subseteq S$. Of course, l^2 is defined as consisting of sequences and, therefore, does not actually consist of "true" individuals. As usual, all that is involved is assuming that S is large enough so that l^2 can be mapped one-one into S; the points of l^2 may then be identified with their images in S.

The orthonormal basis $\{e_n\}$ is a map e of N^+ into l^2. Thus we can regard e as mapping $*N^+$ into $*l^2$. By the Transfer Principle, for all $m,n \in *N^+$,

$$(e_n, e_m) = \delta_m^n = \begin{cases} 0 & \text{for} \quad n \neq m \\ 1 & \text{for} \quad n = m. \end{cases}$$

Also, by the Transfer Principle, for each $a \in *l^2$ there is an internal sequence $\{\alpha_n | n \in *N\}$ such that

$$a = \sum_{n \in *N^+} \alpha_n e_n, \tag{1}$$

where this equation simply means that for every $\varepsilon \in *R^+$, there is an $n_0 \in *N$ such that for $n \in *N$, $n > n_0$,

$$\left\| a - \sum_{k \leq n} \alpha_k e_k \right\| < \varepsilon.$$

For a satisfying (1), we also have

$$\|a\| = \sum_{n \in {}^*N^+} |\alpha_n|^2. \tag{2}$$

Here $\|a\|$ is a hyperreal number such that for every $\varepsilon \in {}^*R^+$, there is an $n_0 \in {}^*N$ such that for $p \in {}^*N$, $p > n_0$,

$$\|a\| - \varepsilon < \sum_{n > p} |\alpha_i|^2 \leq \|a\|.$$

By Theorem 4–1.3, we have

THEOREM 1.5. Let $a \in {}^*l^2$ satisfy (2). Then a is finite if and only if $\sum_{n \in {}^*N^+} |\alpha_i|^2$ is finite.

A somewhat subtler result is

THEOREM 1.6. Let $a \in {}^*l^2$ be finite and satisfy (2). Then a is near standard if and only if $\sum_{\substack{n \in {}^*N^+ \\ n > k}} |\alpha_n|^2 \approx 0$ for all infinite k.

PROOF. First let $a \in l^2$. Then $a = \sum_{n \in N^+} \alpha_n e_n$. Let $r_k = \sum_{\substack{n \in N^+ \\ n > k}} |\alpha_n|^2$, so that $r_k \to 0$ as $k \to \infty$. By Theorem 2–5.1, $r_k \approx 0$ for k infinite. But using the Transfer Principle,

$$r_k = \sum_{\substack{n \in {}^*N^+ \\ n > k}} |\alpha_n|^2$$

for k infinite. And this gives the desired result for $a \in l^2$.

Next let $a \in {}^*l^2$ be near standard, $a \approx b \in l^2$, where

$$b = \sum_{n \in {}^*N^+} \beta_n e_n.$$

Then

$$\|a - b\|^2 = \sum_{n \in {}^*N^+} |\alpha_n - \beta_n|^2 \approx 0.$$

Since $b \in l^2$, $\sum_{\substack{n \in {}^*N^+ \\ n > k}} |\beta_n|^2 \approx 0$ for $k \in {}^*N - N$. Now the triangle inequality for l^2 (Minkowski's inequality) and the Transfer Principle yield:

$$\left(\sum_{\substack{n \in {}^*N^+ \\ n > k}} |\alpha_n|^2 \right)^{1/2} \leq \left(\sum_{\substack{n \in {}^*N^+ \\ n > k}} |\alpha_n - \beta_n|^2 \right)^{1/2} + \left(\sum_{\substack{n \in {}^*N^+ \\ n > k}} |\beta_n|^2 \right)^{1/2},$$

and so $\sum_{\substack{n \in {}^*N^+ \\ n > k}} |\alpha_n|^2 \approx 0$ for k infinite.

For the proof in the converse direction, let $\sum_{n \in {}^*N^+} |\alpha_n|^2$ be finite and suppose that $\sum_{\substack{n \in {}^*N^+ \\ n > k}} |\alpha_n|^2 \approx 0$ for all $k \in {}^*N - N$. Thus α_n is finite for all $n \in {}^*N$. Let $\beta_n = {}^\circ(\alpha_n)$ for $n \in N^+$, and let $b = \sum_{n \in N^+} \beta_n e_n$. We then show that $b \in l^2$ and $a \approx b$.

Since each $\beta_n \approx \alpha_n$ for $n \in N^+$, we have for each $k \in N^+$

$$\sum_{n=1}^{k} |\beta_n|^2 - \sum_{n=1}^{k} |\alpha_n|^2 \approx 0,$$

so that

$$\sum_{n=1}^{k} |\beta_n|^2 \leq \sum_{n=1}^{k} |\alpha_n|^2 + 1.$$

Since a is finite and $\sum_{n=1}^{k} |\alpha_n|^2 \leq \sum_{n \in {}^*N^+} |\alpha_n|^2 = \|a\|$, we have

$$\sum_{n=1}^{k} |\beta_n|^2 \leq \|a\| + 1.$$

Since this bound is independent of k,

$$\sum_{n=1}^{\infty} |\beta_n|^2 < \infty,$$

and hence $b \in l^2$.

Then for any $k \in {}^*N^+$,

$$\begin{aligned} \|a - b\|^2 &= \sum_{n \in {}^*N^+} |\alpha_n - \beta_n|^2 \\ &= \sum_{n=1}^{k} |\alpha_n - \beta_n|^2 + \sum_{\substack{n \in {}^*N \\ n > k}} |\alpha_n - \beta_n|^2. \end{aligned}$$

Now for k finite, $\sum_{n=1}^{k} |\alpha_n - \beta_n|^2 \approx 0$. Hence by the Infinitesimal Prolongation Theorem, the same is true for some infinite k. But for any infinite value of k we have, using Minkowski's inequality once again,

$$\left(\sum_{\substack{n \in {}^*N \\ n > k}} |\alpha_n - \beta_n|^2 \right)^{1/2} \leq \left(\sum_{\substack{n \in {}^*N \\ n > k}} |\alpha_n|^2 \right)^{1/2} + \left(\sum_{\substack{n \in {}^*N \\ n > k}} |\beta_n|^2 \right)^{1/2} \approx 0,$$

using the hypothesis and the first part of this proof. Hence $\|a - b\|^2 \approx 0$, and we are done. ■

PROOF. Let $x,y \in E^\perp$, $\lambda \in C$, and let u be any point of E. Then,

$$(x + y,u) = (x,u) + (y,u) = 0,$$

and

$$(\lambda x,u) = \lambda(x,u) = 0.$$

Thus $x + y \in E^\perp$ and $\lambda x \in E^\perp$, so that $E^\perp$ is a linear subspace of H.

To see that $E^\perp$ is closed, let $x \in {}^*(E^\perp)$, $x \approx y \in H$. Then for any $u \in E$,

$$(y,u) \approx (x,u) = 0.$$

So $(y,u) = 0$, and $y \in E^\perp$. ■

LEMMA 3. For any $x \notin E$, $(x - P_E x) \in E^\perp$.

PROOF. Let $z \in E$, $z \neq 0$, $y = P_E x$, and let $\lambda \in R$, $\lambda \neq 0$. As above, let $\alpha = \|x - y\| > 0$. Then writing "$\mathscr{R}$", "$\mathscr{I}$" for real and imaginary part, respectively,

$$\begin{aligned}\alpha^2 &< \|x - (y + \lambda z)\|^2 \\ &= ((x - y) - \lambda z, (x - y) - \lambda z) \\ &= \|x - y\|^2 + \lambda^2\|z\|^2 - (x - y,\lambda z) - (\lambda z,x - y) \\ &= \alpha^2 + \lambda^2\|z\|^2 - 2\lambda\mathscr{R}(x - y,z).\end{aligned}$$

Thus $A^2\lambda^2 - 2B\lambda > 0$ where $A = \|z\| \neq 0$ and $B = \mathscr{R}(x - y,z)$. But

$$A^2\lambda^2 - 2B\lambda = \left(A\lambda - \frac{B}{A}\right)^2 - \frac{B^2}{A^2} = -\frac{B^2}{A^2}$$

for $\lambda = B/A^2$, and $-B^2/A^2 < 0$ unless $B = 0$. We conclude that for any $z \in E$, if $z \neq 0$, then $\mathscr{R}(x - y,z) = 0$. Hence also,

$$\mathscr{I}(x - y,z) = \mathscr{R}[(-i)(x - y,z)] = \mathscr{R}(x - y,iz) = 0.$$

Finally, $(x - y,z) = 0$ and $x - y \in E^\perp$. ■

LEMMA 4. $y = P_E x$ if and only if $y \in E$ and $x - y \in E^\perp$.

PROOF. Lemma 3 gives the result in one direction.

For the other direction, let $y \in E$, $x - y \in E^\perp$. Then by the Pythagorean Theorem, for all $z \in E$,

$$\begin{aligned}\|x - z\|^2 &= \|(x - y) + (y - z)\|^2 \\ &= \|x - y\|^2 + \|y - z\|^2 \geq \|x - y\|^2.\end{aligned}$$

Hence $y = P_E x$. ■

In particular, the lemma implies that

$$\textit{if } x \in E^\perp, \qquad \textit{then } P_E x = 0.$$

LEMMA 5. P_E is a bounded linear operator, and unless $E = \{0\}$, $\|P_E\| = 1$.

PROOF. Let $y = P_E x$, $y' = P_E x'$. Then by Lemma 4, $y, y' \in E$, $x - y$, $x' - y' \in E^\perp$. Since E and $E^\perp$ are both linear spaces, for any $\lambda \in C$,

$$y + y' \in E, \qquad (x + x') - (y + y') \in E^\perp,$$
$$\lambda y \in E, \qquad \lambda x - \lambda y \in E^\perp.$$

By Lemma 4,

$$P_E(x + x') = y + y' = P_E x + P_E x',$$
$$P_E(\lambda x) = \lambda y = \lambda P_E x.$$

By the Pythagorean theorem,

$$\|x\|^2 = \|P_E x\|^2 + \|x - P_E x\|^2.$$

Hence $\|P_E x\| \leq \|x\|$, which shows that P_E is bounded and $\|P_E\| \leq 1$. Finally, if $E \neq \{0\}$, there exists $u \in E$ with $\|u\| = 1$ (i.e., $u = x/\|x\|$ where $x \in E$, $x \neq 0$). But then since $P_E u = u$, $\|P_E u\| = \|u\|$. Thus $\|P_E\| = 1$.

DEFINITION. If A,B are linear subspaces of H, we write $A \perp B$, and say that A and B are *orthogonal spaces* if $u \perp v$ whenever $u \in A$ and $v \in B$. ■

LEMMA 6. Let E,F be closed linear subspaces of H. Let $E \perp F$, and set

$$G = \{x + y | x \in E \text{ and } y \in F\}.$$

Then G is a closed linear subspace of H, and each element of G is *uniquely* representable as $x + y$ with $x \in E$ and $y \in F$.

PROOF. By hypothesis, $E \subseteq F^\perp$, $F \subseteq E^\perp$. Let $V = I - (P_E + P_F)$. Then $x \in E$ and $y \in F$ implies $V(x + y) = 0$. Conversely, if $Vz = 0$, then $z = P_E z + P_F z$, so that $z \in G$. Hence by Theorem 4–1.13, G is a closed linear subspace of H.

Let $z = x + y$, $x \in E$, $y \in F$. Then by Lemma 4, $x = P_E z$. Similarly, $y = P_F z$. This proves the uniqueness. ■

DEFINITION. Under the hypothesis of Lemma 6, we write $G = E \oplus F$ and call G the *direct sum* of E and F.

Since it is clear that $E_1 \oplus (E_2 \oplus E_3) = (E_1 \oplus E_2) \oplus E_3$ (of course, the notation only makes sense when $E_1 \perp E_2$, $E_2 \perp E_3$ and $E_3 \perp E_1$), the notation $E_1 \oplus E_2 \oplus \cdots \oplus E_n$ for mutually orthogonal closed linear spaces $E_1, \ldots, E_n$ is unambiguous. If $\{E_n | n = 1,2,3, \ldots\}$ is a sequence of mutually orthogonal closed linear subspaces of H, and if $G_n = E_1 \oplus \cdots \oplus E_n$ for each n, then we set

$$E_1 \oplus E_2 \oplus \cdots = \overline{\bigcup_{n=0}^{\infty} G_n},$$

which by Theorem 2.2 is a closed linear subspace of H.

LEMMA 7. $H = E \oplus E^{\perp}$. Hence for each $x \in H$, we have the *unique* decomposition $x = y + z$, $y \in E$, $z \in E^{\perp}$.

PROOF. The decomposition $x = P_E x + (x - P_E x)$ and Lemma 3 shows that $x \in E \oplus E^{\perp}$. The uniqueness then follows from Lemma 6. ■

LEMMA 8. $E^{\perp\perp} = E$.

PROOF. If $x \in E$, then $x \perp y$ for all $y \in E^{\perp}$. Since x is orthogonal to all $y \in E^{\perp}$, we must have $x \in (E^{\perp})^{\perp}$. Thus $E \subseteq E^{\perp\perp}$.

Conversely, if $x \in E^{\perp\perp}$, we write $x = x_1 + y$, $x_1 \in E$, $y \in E^{\perp}$. Since $x \in E^{\perp\perp}$, we have $x \perp y$. Then

$$0 = (x,y) = (x_1 + y,y) = (x_1,y) + \|y\|^2 = \|y\|^2.$$

Therefore, $y = 0$. That is, $x = x_1 \in E$. ■

LEMMA 9. Let $G = E \oplus F$. Then $P_G = P_E + P_F$.

PROOF. Let $u = P_E x$, $v = P_F x$ for $x \in H$. We wish to show that $u + v = P_G x$, that is, using Lemma 4, that $u + v \in G$, $(u + v) - x \in G^{\perp}$. Now, $u + v \in G$ by definition of direct sum. Let t be any element of G and write $t = r + s$, $r \in E$, $s \in F$. It remains to show that $(u + v) - x \perp t$. But this follows easily from the computation:

$$\begin{aligned}(u + v - x, r + s) &= (u + v - x,r) + (u + v - x,s) \\ &= (u - x,r) + (v,r) + (u,s) + (v - x,s) \\ &= 0,\end{aligned}$$

using Lemma 4 and the fact that $E \perp F$. ■

COROLLARY 2.3. $I = P_E + P_{E^{\perp}}$.

PROOF. Clearly $P_H = I$. ■

By an obvious induction, we have

COROLLARY 2.4. If $G = E_1 \oplus E_2 \oplus \cdots \oplus E_n$, then,

$$P_G = P_{E_1} + P_{E_2} + \cdots + P_{E_n}$$

LEMMA 10. $(P_E x,y) = (P_E x,P_E y) = (x,P_E y)$. Thus P_E is Hermitian.

PROOF. Let $y = P_E y + u$, $u \in E^\perp$. Then

$$\begin{aligned}(P_E x, y) &= (P_E x, P_E y) + (P_E x, u) \\ &= (P_E x, P_E y)\end{aligned}$$

Using this,

$$(x, P_E y) = \overline{(P_E y, x)} = \overline{(P_E y, P_E x)} = (P_E x, P_E y). \quad \blacksquare$$

In addition to being Hermitian, P_E is *idempotent*, that is, $P_E^2 = P_E$. (This is because $P_E x \in E$, and, therefore, $P_E(P_E x) = P_E x$). This suggests the

DEFINITION. A bounded linear operator on H is called a *projection* if it is Hermitian and idempotent.

THEOREM 2.5. If P is a projection, then there is a closed linear subspace E of H such that $P = P_E$.

PROOF. Let E be the range of the operator P. If $x \in E$, then $x = Pu$ for some $u \in H$, so that $(I - P)x = Pu - P^2 u = 0$. Conversely, if $(I - P)x = 0$, then $x = Px \in E$. Thus by Theorem 4–1.13, E is a closed linear subspace of H, and P_E is defined.

Now let $y = Px$. By Lemma 4 we need only show that $x - y \perp u$ for all $u \in E$. But using the fact that P is Hermitian and idempotent,

$$\begin{aligned}(x - y, Pv) &= (x, Pv) - (Px, Pv) \\ &= (x, P^2 v) - (Px, Pv) \\ &= (Px, Pv) - (Px, Pv) \\ &= 0. \quad \blacksquare\end{aligned}$$

THEOREM 2.6. Let E,F be closed linear subspaces of H. Then the following are equivalent:

(1) $E \subseteq F$,
(2) $P_E P_F = P_E$,
(3) $\|P_E x\| \leq \|P_F x\|$ for all $x \in H$,
(4) $P_E \leq P_F$.

PROOF. (1) $\Rightarrow$ (2): Let $y = P_E(P_F x)$ for $x \in H$. We prove that $y = P_E x$. By Lemma 4, this amounts to showing that $x - y \in E^\perp$. But by Lemma 3, $P_F x - y \in E^\perp$ and $x - P_F x \in F^\perp \subseteq E^\perp$. Since $E^\perp$ is a linear space,

$$x - y = (x - P_F x) + (P_F x - y) \in E^\perp.$$

(2) $\Rightarrow$ (3): $\|P_E x\| = \|P_E P_F x\| \leq \|P_E\| \cdot \|P_F x\| = \|P_F x\|$.

(3) $\Rightarrow$ (4): We have $(P_E x, P_E x) \leq (P_F x, P_F x)$. By Lemma 10, this can be written: $(P_E x, x) \leq (P_F x, x)$; but by definition, we then have $P_E \leq P_F$.

(4) $\Rightarrow$ (1): Let $x \in E$, that is, let $P_E x = x$. By Corollary 2.3, $I - P_F = P_{F^\perp}$. Hence by Lemma 10,

$$\begin{aligned}\|(I - P_F)x\|^2 &= ((I - P_F)x, (I - P_F)x)\\ &= ((I - P_F)x,x)\\ &= (x,x) - (P_F x,x)\\ &= (P_E x,x) - (P_F x,x) \leq 0.\end{aligned}$$

Hence $(I - P_F)x = 0$, that is, $P_F x = x$ and $x \in F$. ■

LEMMA 11. Let $E \subseteq F$ where E and F are closed linear subspaces of H. Then

(1) $P_E P_F = P_E = P_F P_E$,
(2) $P_F - P_E$ is a projection.

PROOF.

(1) By Theorem 2.6, $P_E P_F = P_E$. Moreover, since by hypothesis, $P_E x \in F$, we have $P_F(P_E x) = P_E x$ for all $x \in H$.
(2) Using Theorem 2.5, we need only verify that $P_F - P_E$ is Hermitian and idempotent. We have

$$\begin{aligned}((P_F - P_E)x,y) &= (P_F x,y) - (P_E x,y)\\ &= (x,P_F y) - (x,P_E y)\\ &= (x,(P_F - P_E)y),\end{aligned}$$

and, using (1),

$$\begin{aligned}(P_F - P_E)^2 &= P_F^2 - P_F P_E - P_E P_F + P_E^2\\ &= P_F - P_E - P_E + P_E = P_F - P_E.\end{aligned}$$ ■

LEMMA 12. Let $\{G_n | n \in N^+\}$ be a sequence of closed linear subspaces of H, such that $G_n \subseteq G_{n+1}$ for all n. Let $G = \bigcup_{n=0}^{\infty} G_n$. Then $P_{G_n} x \to P_G x$ for all $x \in H$.

PROOF. Let $x \in H$, and set $y_n = P_{G_n} x$. By Theorem 2.6 (3), $\|y_n\| \leq \|y_{n+1}\|$ for all n. Also, for each n, $\|y_n\| \leq \|x\|$. Thus the sequence $\{\|y_n\|^2 | n \in N^+\}$ is a monotone increasing bounded sequence of real numbers and hence is a Cauchy sequence. For $m > n$, $P_{G_m} - P_{G_n}$ is a projection by Lemma 11 (2), so that (using Lemma 10)

$$\begin{aligned}\|y_m - y_n\|^2 &= ((P_{G_m} - P_{G_n})x, (P_{G_m} - P_{G_n})x)\\ &= ((P_{G_m} - P_{G_n})x,x)\\ &= (P_{G_m} x,P_{G_m} x) - (P_{G_n} x,P_{G_n} x)\\ &= \|y_m\|^2 - \|y_n\|^2.\end{aligned}$$

Hence $\{y_n | n \in N^+\}$ is a Cauchy sequence in H and thus has a limit, say $y_n \to y$.

It remains to show that $y = P_G x$. To do so, we once again use Lemma 4. Since each $y_n \in G_n \subseteq G$ and G is closed, it follows that $y \in G$. We, therefore, need only show that $x - y \in G^\perp$.

First let $u \in G_k$, and let us compute $(x - y,u)$. We have $(x - y_n,u) = 0$ for all $n \geq k$, since $x - y_n \in G_n^\perp \subseteq G_k^\perp$. Hence $(x - y,u) = 0$. Finally, let $v \in G$. Then for some sequence $\{u_k | k \in N^+\}$, we have $u_k \to v$, and each $u_k \in \bigcup_{n=0}^{k} G_n$. Then $(x - y,u_k) = 0$, and, therefore, $(x - y,v) = 0$. Thus we have shown that $x - y \in G^\perp$. ■

LEMMA 13. Let $\{E_n | n \in N^+\}$ be a sequence of closed linear subspaces of H. Let $G = E_1 \oplus E_2 \oplus E_3 \oplus \cdots$. Then for each $x \in H$, $P_G x = \sum_{n=1}^{\infty} P_{E_n} x$.

PROOF. Let $G_n = E_1 \oplus \cdots \oplus E_n$. Using Corollary 2.4 $P_G x = P_{E_1} x + \cdots + P_{E_n} x$ for all $x \in H$. The result follows from Lemma 12.

LEMMA 14. Let $\{Q_n | n \in N^+\}$ be a sequence of projections on H such that $Q_{n+1} \leq Q_n$ for each n. Then the sequence $\{Q_n x | n \in N^+\}$ has a limit for each $x \in H$.

PROOF. Let $Q_n = P_{K_n}$ so that by Theorem 2.6, $K_n \subseteq K_{n+1}$. Let $G_n = K_n^\perp$ for each n, so that $I - Q_n = P_{G_n}$. Now, $G_n \subseteq G_{n+1}$. Hence by Lemma 12, for each $x \in H$, there is a $y \in H$ such that

$$(I - Q_n)x \to y.$$

But then

$$Q_n x \to x - y. \quad ■$$

Next we develop the classical Gram–Schmidt process for orthonormalizing a basis. Let $\{t_1, \ldots, t_k\}$ be an orthonormal sequence in H, let $E = \operatorname{span}(t_1, \ldots, t_k)$, and let $u \in H - E$. By Theorem 2.1, E is closed. Hence we may set $v = u - P_E u$, so that v is orthogonal to $t_1, \ldots, t_k$. Finally, letting $t_{k+1} = v/\|v\|$, we have

(1) $\{t_1, \ldots, t_k, t_{k+1}\}$ is orthonormal,
(2) $\operatorname{span}(t_1, \ldots, t_k, t_{k+1}) = \operatorname{span}(t_1, \ldots, t_k, u)$.

Since for a single point $u \in E$, $u \neq 0$, it is obvious that $\operatorname{span}(u) = \operatorname{span}(u/\|u\|)$, we have proved by induction:

THEOREM 2.7 (GRAM–SCHMIDT PROCESS). Let $\{u_n | n \in I\}$ be a (finite or infinite) sequence of linearly independent points of H. Then there is an orthonormal sequence $\{t_n | n \in I\}$ such that for each $k \in I$,

$$\text{span}(u_1, \ldots, u_k) = \text{span}(t_1, \ldots, t_k).$$

COROLLARY 2.8. A finite dimensional subspace of H is closed.

PROOF. The proof is immediate from Theorems 2.1 and 2.7. ■

LEMMA 15. Let $E = \text{span}(t_1, \ldots, t_k)$ where $\{t_1, \ldots, t_k\}$ is an orthonormal sequence in H. Then for all $x \in H$,

$$P_E x = \sum_{i=1}^{k} (x,t_i)t_i.$$

PROOF. Let $y = \sum_{i=1}^{k} (x,t_i)t_i$. Since y is obviously in E, we need only show (by Lemma 4) that $x - y \in E^\perp$. But this will follow at once if we can show that

$$(x - y,t_j) = 0, \qquad j = 1,2, \ldots, k.$$

And we have

$$\begin{aligned}\left(x - \sum_{i=1}^{k} (x,t_i)t_i,t_j\right) &= (x,t_j) - \sum_{i=1}^{k} (x,t_i)(t_i,t_j) \\ &= (x,t_j) - (x,t_j) \\ &= 0. \ ■\end{aligned}$$

Returning to the special case $H = l^2$ of Section 1, we note that P_n is just P_{H_n}. For, let $x \in l^2$, say

$$x = \sum_{j=1}^{\infty} a_j e_j, \qquad a_j \in C.$$

Then by Lemma 15,

$$\begin{aligned}P_{H_n}(x) &= \sum_{i=1}^{n} (x,e_i)e_i \\ &= \lim_{m\to\infty} \sum_{i=1}^{n} \left(\sum_{j=1}^{m} a_j(e_j,e_i)\right) e_i \\ &= \lim_{m\to\infty} \sum_{j=1}^{m} a_j \sum_{i=1}^{n} (e_j,e_i)e_i.\end{aligned}$$

Now,

$$\sum_{i=1}^{n} (e_j,e_i)e_i = \begin{cases} e_j & \text{if } j \le n \\ 0 & \text{if } j > n. \end{cases}$$

Hence

$$P_{H_n}(x) = \sum_{j=1}^{n} a_j e_j = P_n x.$$

We now write $\mathscr{E}$ for the set of finite dimensional subspaces of H. Since $\mathscr{E} \subseteq \mathscr{P}(H)$, certainly $\mathscr{E} \in U$. By Theorem 1–7.12, $^*\mathscr{E} \subseteq {^*\mathscr{P}}(H)$. We write dim for the function mapping $\mathscr{E}$ into N such that $\dim(E)$ is the dimension of the space E. Then, *dim maps $^*\mathscr{E}$ into $^*N$. By the Transfer Principle, if $E \in {^*\mathscr{E}}$ and $^*\dim(E) = \nu$, then there is an internal sequence $\{t_i | 1 \le i \le \nu\}$ (the basis of E) such that $t_i \in {^*H}$ for $1 \le i \le \nu$ and

$$E = \left\{ \sum_{i=1}^{\nu} \alpha_i t_i \middle| \alpha_i \in {^*C} \right\}. \tag{‡}$$

Also, if E is a subset of $^*H$ and $\{t_i | 1 \le i \le \nu\}$ is an internal sequence of points of $^*H$ for which (‡) holds, then $E \in {^*\mathscr{E}}$ and $^*\dim(E) \le \nu$. Thus if $E \in \mathscr{E}$, then $E \in {^*\mathscr{E}}$ and $\dim(E) = {^*\dim}(E)$. Hence we are able to omit the * as usual, writing simply $\dim(E)$ for all $E \in {^*\mathscr{E}}$.

Since finite dimensional subspaces of H are closed, $P_E x$ is defined for all $E \in \mathscr{E}$, $x \in H$. Regarding P as a map of $\mathscr{E} \times H$ into H, $^*P$ maps $^*\mathscr{E} \times {^*H}$ into $^*H$ so we may speak of $^*P_E x$ for all $E \in {^*\mathscr{E}}$, $x \in {^*H}$. Using Lemma 4 and the Transfer Principle, we see that $^*P_E x = P_E x$ for $x \in H$, $E \in \mathscr{E}$, so once again, we can harmlessly omit the *.

Using the Transfer Principle, various of the properties of orthogonal projections are easily seen to hold for P_E when $E \in {^*\mathscr{E}}$. For example, $P_E x = x$ for all $x \in {^*H}$ and $E \in {^*\mathscr{E}}$. Also, $\|P_E\| = 1$.

Returning finally to the case $H = l^2$, we note that $H_n \subseteq \mathscr{E}$ and $\dim(H_n) = n$ for all $n \in N$. Thus for $n \in {^*N}$, we have $H_n \subseteq {^*\mathscr{E}}$ and $\dim(H_n) = n$.

3. THE BERNSTEIN–ROBINSON THEOREM

In this section, we introduce a technique for applying results of finite dimensional linear algebra to infinite dimensional spaces that is used in the remainder of this chapter. The idea is to find a space $E \in {^*\mathscr{E}}$ that approximates the given infinite dimensional space. Then the Transfer Principle permits us to apply the results of the finite dimensional theory to the approximating space, and hence ultimately to the given infinite dimensional space.

In this section we show how this method was used by Bernstein and Robinson to solve a problem that had been open for many years. First we introduce some terminology:

DEFINITION. T is called *polynomially compact* if T is a bounded linear operator and for some polynomial

$$p(\lambda) = c_0 + c_1\lambda + \cdots + c_n\lambda^n, \; c_0, \ldots, c_n \in C,$$

the operator $p(T)$ is compact.

DEFINITION. The subspace E of the unitary space H is *invariant* for the operator T if $T[E] \subseteq E$.

In the 1930s J. von Neumann had proved (but not published) the fact that a compact operator on Hilbert space has a nontrivial (i.e., neither $\{0\}$ nor the whole space) invariant subspace. The question of which noncompact operators possess nontrivial invariant subspaces was raised, but there was almost no progress. Then, in 1966 nonstandard methods were used to obtain the

THEOREM 3.1 (BERNSTEIN–ROBINSON THEOREM). Let T be a polynomially compact operator on l^2. Then there is a closed linear subspace of l^2, other than $\{0\}$ or l^2, which is invariant for T.

We begin our discussion[1] with some general considerations. In this section, we consider only $H = l^2$. Let $\{e_i | i \in N^+\}$ be some orthonormal basis for l^2. (Later we shall want to specify this basis.) Then if T is a linear operator on l^2, we may write

$$Te_k = \sum_{j=1}^{\infty} a_{jk} e_j, \qquad k = 1,2,3, \ldots .$$

The array $[a_{jk}]$ (which is, of course, just a map of $N^+ \times N^+$ into C) is then called the *matrix* of T (with respect to the given basis). As usual, we can think of a_{jk} as defined for $j,k \in {}^*N^+$ with values in ${}^*C$.

LEMMA 1. Let $[a_{jk}]$ be the matrix of a compact operator T. Then $a_{jk} \approx 0$ for $j \in {}^*N - N$, and all $k \in N^+$.

PROOF. By the Transfer Principle,

$$Te_k = \sum_{i \in {}^*N^+} a_{ik} e_i .$$

Since T is compact and e_k is finite, Te_k is near standard (Theorem 4–2.1). Thus $Te_k \approx y \in l^2$ where, say, $y = \sum_{i \in {}^*N^+} y_i e_i$. Let

$$\begin{aligned} \xi &= \|Te_k - y\|^2 \\ &= \left\| \sum_{i \in {}^*N^+} (a_{ik} - y_i) e_i \right\|^2 \\ &= \sum_{i \in {}^*N^+} |a_{ik} - y_i|^2, \end{aligned}$$

using the Pythagorean theorem. Since $Te_k \approx y$, $\xi \approx 0$. Choose any $j \in {}^*N - N$. Clearly,

$$|a_{jk} - y_j|^2 \leq \xi,$$

and hence

$$|a_{jk} - y_j| \leq \xi^{1/2}.$$

[1] We follow Bernstein [3].

Then

$$|a_{jk}| = |a_{jk} - y_j + y_j| \\ \leq \xi^{1/2} + |y_j|.$$

Since $y \in l^2$, $\sum_{n=1}^{\infty} |y_n|^2$ converges so that $y_n \to 0$, and hence $|y_j| \approx 0$. We conclude that

$$|a_{jk}| \approx 0. \quad \blacksquare$$

DEFINITION. A matrix $[a_{jk}]$ is called *almost superdiagonal* if $a_{jk} = 0$ for all $j > k + 1$.

Let $[a_{jk}]$ be almost superdiagonal. Then we define recursively

$$a_{jk}^{(1)} = a_{jk},$$

$$a_{jk}^{(n+1)} = \sum_{i=1}^{\infty} a_{ji}^{(n)} a_{ik} = \sum_{i \leq k+1} a_{ji}^{(n)} a_{ik}.$$

LEMMA 2. Let the matrix of the bounded linear operator T be the almost superdiagonal matrix $[a_{jk}]$. Then the matrix of T^n is $[a_{jk}^{(n)}]$.

PROOF. The proof is by induction on n. For $n = 1$ the result is immediate. Suppose the result holds for n. Let $[b_{jk}]$ be the matrix of T^{n+1}. Then

$$\begin{aligned} \sum_{j=1}^{\infty} b_{jk} e_j &= T^{n+1} e_k \\ &= T^n(Te_k) \\ &= T^n\left(\sum_{i=1}^{\infty} a_{ik} e_i\right) \\ &= T^n\left(\sum_{i \leq k+1} a_{ik} e_i\right) \\ &= \sum_{i \leq k+1} a_{ik}(T^n e_i) \\ &= \sum_{i \leq k+1} a_{ik} \sum_{j=1}^{\infty} a_{ji}^{(n)} e_j \\ &= \sum_{j=1}^{\infty} \left(\sum_{i \leq k+1} a_{ji}^{(n)} a_{ik}\right) e_j \\ &= \sum_{j=1}^{\infty} a_{jk}^{(n+1)} e_j. \quad \blacksquare \end{aligned}$$

LEMMA 3. Let $[a_{jk}]$ be almost superdiagonal. Then for $s,m,n > 0$,

$$a^{(s)}_{n+m,n} = 0 \quad \text{if} \quad s < m, \tag{1}$$

$$a^{(m)}_{n+m,n} = \prod_{i=0}^{m-1} a_{n+i+1,\,n+i}. \tag{2}$$

PROOF. We first prove (1) by induction on s. For $s = 1$, $m \geq 2$. Hence since $[a_{jk}]$ is almost superdiagonal, $a^{(1)}_{n+m,n} = a_{n+m,n} = 0$.

Assume that (1) holds for superscripts $<s$. By the induction hypothesis,

$$a^{(s-1)}_{n+m,n} = 0$$

for $s - 1 < m$ and all m,n. But

$$a^{(s)}_{n+m,n} = \sum_{i \leq n+1} a^{(s-1)}_{n+m,i}\, a_{i,n}.$$

But for $i \leq n + 1$ and $m > s$, by the induction hypothesis, all terms of the sum are 0, and hence so is $a^{(s)}_{n+m,n}$.

Next we prove (2) by induction on m. For $m = 1$ the result is trivial. Suppose that (2) holds for $m - 1$, that is, that

$$a^{(m-1)}_{n+m-1,n} = \prod_{i=0}^{m-2} a_{n+i+1,\,n+i}$$

for all n. Using (1), $a^{(m-1)}_{n+m,i} = 0$ for $i \leq n$. So,

$$\begin{aligned} a^{(m)}_{n+m,n} &= \sum_{i \leq n+1} a^{(m-1)}_{n+m,i}\, a_{in} \\ &= a^{(m-1)}_{n+m,\,n+1}\, a_{n+1,n} \\ &= \left(\prod_{i=0}^{m-2} a_{n+i+2,\,n+i+1}\right) \cdot a_{n+1,n} \\ &= \prod_{i=0}^{m-1} a_{n+i+1,\,n+i}. \end{aligned}$$

LEMMA 4. Let T be a polynomially compact operator on l^2 whose matrix $[a_{jk}]$ is almost superdiagonal. Then for some infinite integer ν, $a_{\nu+1,\nu} \approx 0$.

PROOF. Let $p(T) = \alpha_0 + \alpha_1 T + \cdots + \alpha_K T^K$ be compact and have the matrix $[b_{jk}]$. Then using Lemma 2,

$$b_{n+m,n} = 0 + \alpha_1 a^{(1)}_{n+m,n} + \alpha_2 a^{(2)}_{n+m,n} + \cdots + \alpha_K a^{(K)}_{n+m,n}.$$

Using (1) and then (2) of Lemma 3,

$$b_{n+m,n} = \alpha_m a^{(m)}_{n+m,n} = \alpha_m \prod_{i=0}^{m-1} a_{n+i+1,\,n+i}.$$

Since $p(T)$ is compact, Lemma 1 implies that $b_{n+m,n} \approx 0$ for infinite n. Choosing such an n, one of the numbers $a_{n+i+1,n+i}$, $i = 0, \ldots, m-1$ must be infinitesimal. This gives the result. ■

DEFINITION. For $E \in {}^*\mathscr{E}$, let ${}^\circ E$ be the set of $x \in l^2$ such that $x \approx y$ for some $y \in E$.

LEMMA 5. ${}^\circ E$ is a closed linear subspace of l^2.

PROOF. Let $x,y \in {}^\circ E$ and $\lambda \in C$. Since $x \approx x' \in E$ and $y \approx y' \in E$, $x + y \approx x' + y' \in E$; so $x + y \in {}^\circ E$. Also, $\lambda x \approx \lambda x' \in E$, so that $\lambda x \in {}^\circ E$.

To see that ${}^\circ E$ is closed, we let x be a limit point of ${}^\circ E$ and show that $x \in {}^\circ E$. Let $n \in N^+$. Then there exists $x' \in {}^\circ E$ such that $\|x - x'\| < 1/2n$. Now $x' \approx y$ for some point $y \in E$, so that certainly $\|x' - y\| < 1/2n$. Then $\|x - y\| < 1/n$. Let

$$K = \{n \in {}^*N^+ | * \models (\exists \mathbf{y} \in \mathbf{E})(\|\mathbf{x} - \mathbf{y}\| < \mathbf{1/n})\}.$$

By the above, $N^+ \subseteq K$. Since K is internal, there must be an infinite $\nu \in K$. Thus there exists a $y \in E$ such that $\|x - y\| < 1/\nu \approx 0$. That is, $x \in {}^\circ E$. ■

We make use of the following fundamental result from finite dimensional linear algebra (cf., for example, Halmos [5], Section 56).

INVARIANCE THEOREM. Let E be a finite dimensional linear space of dimension m and T a linear operator on E. Then there exists a chain

$$\{0\} = E_0 \subset E_1 \subset E_2 \subset \cdots E_m = E,$$

where $\dim(E_i) = i$ and each E_i is invariant for T.

This result is equivalent to the existence of a basis in terms of which T has a "triangular" matrix. Our plan is to use the Transfer Principle and Lemma 5 to obtain invariant subspaces in l^2.

LEMMA 6. For $x \in l^2$ and $E \in {}^*\mathscr{E}$, $x \in {}^\circ E$ if and only if $x \approx P_E x$.

PROOF. If $x \approx P_E x \in E$, then, of course, $x \in {}^\circ E$. Conversely, if $x \in {}^\circ E$, then $x \approx y$ for some $y \in E$. By the definition of P_E and the Transfer Principle,

$$\|x - P_E x\| \leq \|x - y\| \approx 0. \quad ■$$

Let T be some given bounded linear operator on l^2. Let $\nu \in {}^*N - N$, let H_ν, P_ν be as in Section 1, let $T_\nu = P_\nu T P_\nu$, and let $\tilde{T}$ be the restriction of the operator T_ν to H_ν. Thus T_ν is a linear operator on ${}^*l^2$, $\tilde{T}$ is a linear operator on H_ν, and

$$\|\tilde{T}\| \leq \|T_\nu\| \leq \|P_\nu T P_\nu\| \leq \|T\|.$$

Since $H_\nu \in {}^*\mathscr{E}$, for any internal linear subspace E of H_ν, we have $E \in {}^*\mathscr{E}$. ■

LEMMA 7. If E is a linear subspace of H_v and $T_v[E] \subseteq E$, then we have $T[{}^\circ E] \subseteq {}^\circ E$.

PROOF. Let $x \in {}^\circ E$; we wish to show that $Tx \in {}^\circ E$. By Lemma 6, $P_E X \approx x$. Since T is continuous, $TP_E x \approx Tx$, and hence $TP_E x$ is near standard. By Theorem 1.8, $P_v(TP_E x) \approx TP_E x$.

Now,

$$T_v P_E x = P_v T P_E x \approx TP_E x \approx Tx.$$

But $P_E x \in E$, and hence by hypothesis $T_v P_E x \in E$. Thus Tx is infinitesimally close to a point of E (namely, $T_v P_E x$) and hence by definition, $Tx \in {}^\circ E$. ■

LEMMA 8. Let $E,F \in {}^*\mathscr{E}$ such that $E \subseteq F$ and $\dim(F) = \dim(E) + 1$. Then ${}^\circ E \subseteq {}^\circ F$, and for any pair $x,y \in {}^\circ F$ there is a $\lambda \in C$ and a $z \in {}^\circ E$ such that either $x = \lambda y + z$ or $y = \lambda x + z$.

PROOF. If $x \in {}^\circ E$, then $x \approx y$ for some $y \in E$. Since $E \subseteq F$, $y \in F$, and hence $x \in {}^\circ F$. Thus ${}^\circ E \subseteq {}^\circ F$.

Let $x,y \in {}^\circ F$. Then $x \approx x'$, $y \approx y'$ for $x',y' \in F$. Since $E,F \in {}^*\mathscr{E}$, and since $\dim(F) = \dim(E) + 1$, by the Transfer Principle,

$$y' = \lambda x' + z \quad (\text{or } x' = \lambda y' + z)$$

with $\lambda \in {}^*C$, $z \in E$. We consider two cases.

CASE 1

λ is finite.

Since x' and y' and λ are near standard, so is

$$z = y' - \lambda x' \quad (\text{or } z = x' - \lambda y').$$

Thus

$$\begin{aligned} {}^\circ z &= {}^\circ y' - {}^\circ(\lambda x') \quad (\text{or } {}^\circ z = {}^\circ x' - {}^\circ(\lambda y')) \\ &= y - ({}^\circ\lambda)x \quad (\text{or } {}^\circ z = x - ({}^\circ\lambda)y) \end{aligned}$$

Hence

$$y = {}^\circ\lambda x + {}^\circ z \quad (\text{or } x = {}^\circ\lambda y + {}^\circ z)$$

with ${}^\circ z \in {}^\circ E$.

CASE 2

λ is infinite.

Dividing by λ, we have

$$x' = \frac{1}{\lambda} y' - \frac{1}{\lambda} z \quad \left(\text{or } y' = \frac{1}{\lambda} x' - \frac{1}{\lambda} z\right)$$

But because $z \in E$, $(1/\lambda)z$ is also in E. Hence we are back to the first case. ■

We are now ready to prove the Bernstein–Robinson theorem.

Fix some $x \in l^2$ where $\|x\| = 1$, and consider the set $A = \{x, Tx, T^2x, \ldots\}$. We begin by eliminating two trivial special cases.

CASE 1

A *is not a linearly independent set.*

Let k be the least integer such that

$$T^k x = \alpha_1 x + \alpha_2 Tx + \cdots + \alpha_k T^{k-1}x, \quad \alpha_1, \ldots, \alpha_k \in C. \tag{1}$$

Then we set

$$E = \operatorname{span}(x, Tx, \ldots, T^{k-1}x)$$

and claim that E meets the requirements of the Bernstein–Robinson theorem. We have $E \neq \{0\}$ because $x \in E$, $\|x\| = 1$. Since E is finite dimensional, it is closed (Corollary 2.8) and $\neq l^2$. Finally, if $a \in E$, say

$$a = \beta_1 x + \beta_2 Tx + \cdots + \beta_k T^{k-1}x, \tag{2}$$

then,

$$Ta = \beta_1 Tx + \beta_2 T^2 x + \cdots + \beta_k T^k x, \tag{3}$$

so that by (1), $Ta \in E$.

Having disposed of Case 1, we may assume that A is linearly independent. Let $G_n = \operatorname{span}(x, Tx, \ldots, Tx^{n-1})$, and let $E = \bigcup_{n=1}^{\infty} G_n$, $F = \bar{E}$. Then by Theorem 2.2, F is a closed linear subspace of l^2.

CASE 2

$F \neq l^2$.

Clearly, $F \neq \{0\}$. We now show that F is invariant for T. If $a \in E$, then we may write an equation of the form (2), and then (3) shows that $Ta \in E$; thus E is invariant for T. Finally, let $a \in F$. Then there must be a sequence $\{a_n | n \in N^+\}$ of points of E such that $a_n \to a$. But then $Ta_n \to Ta$, and, therefore, $Ta \in F$.

There now remains

CASE 3

$F = l^2$.

Since the set A is linearly independent, the points

$$f_i = T^{i-1}x, \qquad i = 1,2,3, \ldots$$

form a basis for l^2. Applying the Gram–Schmidt process (Theorem 2.7) to the points $\{f_i\}$, we obtain an orthonormal basis $\{e_i\}$ of l^2 such that for each $n \in N^+$,

$$\operatorname{span}\{e_1, \ldots, e_n\} = \operatorname{span}\{f_1, \ldots, f_n\}.$$

It is this orthonormal basis that we use in completing the proof.

Thus for each $k \in N^+$,

$$e_k \in \operatorname{span}\{f_1, \ldots, f_k\},$$

so that

$$Te_k \in \operatorname{span}(f_2, \ldots, f_{k+1}) \subseteq \operatorname{span}(e_1, \ldots, e_k, e_{k+1}).$$

Let $[a_{jk}]$ be the matrix of T with respect to the orthonormal basis $\{e_1, e_2, \ldots\}$. Then we have

$$Te_k = \sum_{j=1}^{\infty} a_{jk} e_j = \sum_{j=1}^{k+1} a_{jk} e_j,$$

so that $[a_{ij}]$ is almost superdiagonal. Using the hypothesis that T is polynomially compact and Lemma 4, we conclude that for some infinite ν (which we hereby fix), $a_{\nu+1,\nu} \approx 0$.

Next we show that $\tilde{T}$ is a good approximation to T:

LEMMA 9. If $\xi \in H_\nu$ and ξ is finite, then $T\xi \approx \tilde{T}\xi$.

PROOF. Since $H_\nu = \operatorname{span}(e_1, \ldots, e_\nu)$, we have

$$\xi = \sum_{k=1}^{\nu} \alpha_k e_k, \qquad \alpha_k \in {}^*C.$$

By the Pythagorean theorem,

$$\|\xi\|^2 = \sum_{k=1}^{\nu} |\alpha_k|^2 \geq |\alpha_\nu|^2;$$

hence α_ν is finite. Then since $[a_{jk}]$ is almost superdiagonal,

$$\begin{aligned} T\xi = \sum_{k=1}^{\nu} \alpha_k Te_k &= \sum_{k=1}^{\nu} \alpha_k \sum_{j=1}^{k+1} a_{jk} e_j \\ &= \sum_{k=1}^{\nu} \alpha_k \sum_{j=1}^{\nu+1} a_{jk} e_j \\ &= \sum_{j=1}^{\nu+1} \left(\sum_{k=1}^{\nu} \alpha_k a_{jk} \right) e_j. \end{aligned}$$

Hence

$$P_\nu T\xi = \sum_{j=1}^{\nu} \left(\sum_{k=1}^{\nu} \alpha_k a_{jk} \right) e_j.$$

Thus

$$T\xi = P_\nu T\xi + \sum_{k=1}^{\nu} \alpha_k a_{\nu+1,k} e_{\nu+1}$$
$$= P_\nu T\xi + \alpha_\nu a_{\nu+1,\nu} e_{\nu+1},$$

since the matrix is almost superdiagonal. Therefore,

$$\|T\xi - \tilde{T}\xi\| = \|T\xi - T_\nu \xi\| = |\alpha_\nu| \cdot |a_{\nu+1,\nu}| \approx 0,$$

using the fact that α_ν is finite. ■

LEMMA 10. If ξ is finite and $\xi \in H_\nu$, then $T^n \xi \approx \tilde{T}^n \xi$ for all $n \in N^+$.

PROOF. The case $n = 1$ was Lemma 9. Suppose the result holds for $n - 1$. Now $\|\tilde{T}^{n-1}\xi\| \leq \|T\|^{n-1}\|\xi\|$ is finite. Therefore, using the induction hypothesis, the continuity of T, and Lemma 9,

$$\begin{aligned} T^n \xi &= T(T^{n-1}\xi) \\ &\approx T(\tilde{T}^{n-1}\xi) \\ &\approx \tilde{T}(\tilde{T}^{n-1}\xi) \\ &= \tilde{T}^n \xi. \quad ■ \end{aligned}$$

Now, let $p(\lambda)$ be the given polynomial such that $p(T)$ is compact.

LEMMA 11. Let ξ be finite and $\xi \in H_\nu$. Then $p(T)\xi \approx p(\tilde{T})\xi$.

PROOF. The proof is immediate from Lemma 10. ■

LEMMA 12. $p(T_\nu)$ is microcontinuous on $^*l^2$.

PROOF. Let $p(\lambda) = \sum_{i=0}^{K} \alpha_i \lambda^i$. Then using Theorem 4–1.11,

$$\begin{aligned} \|p(T_\nu)\| &\leq \sum_{i=0}^{K} |\alpha_i| \cdot \|T_\nu\|^i \\ &\leq \sum_{i=0}^{K} |\alpha_i| \cdot \|T\|^i = M, \end{aligned}$$

where M is a positive real number. Thus if $u \approx v$, $u,v \in {}^*l^2$, we have

$$\begin{aligned} \|p(T_\nu)u - p(T_\nu)v\| &\leq \|p(T_\nu)\| \cdot \|u - v\| \\ &\leq M \cdot \|u - v\| \approx 0. \quad ■ \end{aligned}$$

LEMMA 13. $p(T_\nu)x \not\approx 0$.

PROOF. $p(T)x \neq 0$, since otherwise the points $T^i x$ would be linearly dependent. By Theorem 1.8 $P_\nu x \approx x$. Thus $p(T)P_\nu x \approx p(T)x$, since $p(T)$ is a bounded linear operator (Theorem 4–1.11), and $p(T_\nu)P_\nu x \approx p(T_\nu)x$ by Lemma 12. By Lemma 11, $p(T_\nu)x \approx p(T)x$, that is, $^\circ(p(T_\nu)x) = p(T)x \neq 0$. ∎

Now we use the Invariance Theorem from finite dimensional linear algebra quoted above, and the Transfer principle to obtain an internal sequence

$$\{0\} = E_0 \subset E_1 \subset \cdots \subset E_\nu = H_\nu,$$

where $\dim(E_i) = i$ and

$$\tilde{T}[E_i] \subseteq E_i \quad \text{for} \quad i = 0,1,\ldots,\nu.$$

Since $H_\nu \in \mathscr{E}$, each $E_i \in \mathscr{E}$ and the projection P_{E_i} is defined. (Recall the discussion at the end of Section 2.) We next define the internal sequence $\{r_j | j = 0,1,\ldots,\nu\}$ of hyperreal numbers by

$$r_j = \|p(T_\nu)x - p(T_\nu)P_{E_j}x\|.$$

Now, since $E_0 = \{0\}$, $P_{E_0}x = 0$. Thus $r_0 = \|p(T_\nu)x\|$, and by Lemma 13, there is a positive real number r such that $r_0 > r$. On the other hand, $P_{E_\nu} = P_{H_\nu} = P_\nu$, so that

$$r_\nu = \|p(T_\nu)x - p(T_\nu)P_\nu x\| \approx 0$$

by Lemma 12. In particular, $r_\nu < \frac{1}{2}r$. By Theorem 1–8.3, there is a least μ such that $r_\mu < \frac{1}{2}r$, and hence $r_{\mu-1} \geq \frac{1}{2}r$.

Now, by Lemmas 5 and 7, the spaces $^\circ E_j$, $j = 0,1,\ldots,\nu$ are closed linear subspaces of l^2 that are invariant for T. Hence it remains only to show that at least one of them is nontrivial. In fact, we show that either $^\circ E_{\mu-1}$ or $^\circ E_\mu$ is actually nontrivial.

LEMMA 14. $x \notin {}^\circ E_{\mu-1}$, so that $^\circ E_{\mu-1} \neq l^2$.

PROOF. Suppose $x \in {}^\circ E_{\mu-1}$. Then by Lemma 6, $x \approx P_{E_{\mu-1}}x$. But then by Lemma 12, we would have

$$r_{\mu-1} = \|p(T_\nu)x - p(T_\nu)P_{E_{\mu-1}}x\| \approx 0,$$

which is impossible because $r_{\mu-1} \geq \frac{1}{2}r$. ∎

LEMMA 15. $^\circ E_\mu \neq \{0\}$.

PROOF. Let $y = p(T_\nu)P_{E_\mu}x$. Since E_μ is invariant for $\tilde{T}$ (and hence also for $p(\tilde{T})$), $y \in E_\mu$. By Lemma 11, $y \approx p(T)P_{E_\mu}x$. Now, since $p(T)$ is compact and $P_{E_\mu}x$ is finite, it follows (Theorem 4–2.1) that $p(T)P_{E_\mu}x$ is near standard, and so is y. Let $y \approx z \in l^2$; then $z \in {}^\circ E_\mu$. To prove the lemma, it suffices to show that $z \neq 0$.

If, to the contrary, $z = 0$, then $y \approx 0$. But then $r_\mu \approx \|p(T_\nu)x\| > r$, which is impossible since $r_\mu < \frac{1}{2}r$. ■

By Lemmas 14 and 15, if both ${}^\circ E_{\mu-1}$, ${}^\circ E_\mu$ are trivial, we must have

$$ {}^\circ E_{\mu-1} = \{0\}, \qquad {}^\circ E_\mu = l^2. $$

But $\dim(E_\mu) = \mu = \dim(E_{\mu-1}) + 1$. This would imply by Lemma 8 that for any pair of points $x,y \in l^2$, we have $y = \lambda x$ or $x = \lambda y$ for some $\lambda \in C$. But this is, of course, false.

We conclude that either ${}^\circ E_{\mu-1}$ or ${}^\circ E_\mu$ is nontrivial, and the proof of the Bernstein–Robinson theorem is complete. ■

4. THE SPECTRAL THEOREM FOR COMPACT HERMITIAN OPERATORS

In this section, the technique of finite dimensional approximation from above, which was used in Section 3 to construct an invariant subspace, is used to obtain a classical result: the spectral resolution of a compact Hermitian operator on a Hilbert space. This time we will begin with an arbitrary given Hilbert space H and use the Concurrence Theorem to obtain a space $E \in {}^*\mathscr{E}$ (recall the discussion at the end of Section 2) such that

$$ H \subset E \subset {}^*H. $$

That is, H will be embedded in a space that behaves as though it were finite dimensional.

Thus let

$$ r = \{\langle x,E\rangle \mid x \in E,\ E \in \mathscr{E}\}. $$

Since for any $x \in H$, we have $\langle x,\operatorname{span}(x)\rangle \in r$, it follows that $\operatorname{dom}(r) = H$. Moreover, if $x_1, \ldots, x_n \in H$, then $\langle x_i,\operatorname{span}(x_1, \ldots, x_n)\rangle \in r$ for $i = 1,2, \ldots, n$; so, r is concurrent. By the Concurrence Theorem, there is an E such that

$$ \langle x,E\rangle \in {}^*r $$

for all $x \in H$. By the Transfer Principle, we have $E \in {}^*\mathscr{E}$, and for each x in H, x must also be in E, that is, $H \subseteq E$.

For the remainder of this chapter, E is the space just obtained, and we write $P = P_E$.

LEMMA 1. If $x \in {}^*H$ is near standard, then $Px \approx x$.

PROOF. Let $x = y + \delta$ where $y \in H$ and $\delta \approx 0$. Then $Px = Py + P\delta = y + P\delta \approx y \approx x$, since $y \in E$ and $\|P\delta\| \le \|\delta\| \approx 0$. ■

Let T be some bounded linear operator on H; we write T' for the operator PT restricted to E. Since $Px = x$ for $x \in E$, T' can equivalently be described as PTP restricted to E.

LEMMA 2. $\|T'\| \leq \|T\|$. For $x \in H$, $T'x = Tx$.

PROOF.

$$\|T'\| \leq \|PT\| \leq \|P\| \cdot \|T\| = \|T\|.$$

If $x \in H$, then $Tx \in H \subseteq E$. Hence $T'x = P(Tx) = Tx$. ∎

LEMMA 3. If T is Hermitian on H, then T' is Hermitian on E.

PROOF. For $x,y \in E$,

$$\begin{aligned}(T'x,y) &= (PTPx,y)\\ &= (TPx,Py)\\ &= (TPx,y)\\ &= (Px,Ty)\\ &= (x,PTy)\\ &= (x,T'y).\end{aligned}$$ ∎

For the rest of *this section* $T \neq 0$ is some given *compact Hermitian* operator on a Hilbert space H; thus T' is a Hermitian operator on E, and $\|T'\| \leq \|T\|$.

LEMMA 4. If $x \in E$ is finite, then $T'x$ is near standard.

PROOF. Since T is compact, by Theorem 4–2.1, Tx is near standard. Thus by Lemma 1, $T'x = PTx \approx Tx$, and hence $T'x$ is near standard. ∎

The *eigenvalue equation* for an operator U on a unitary space is

$$(\lambda I - U)x = 0.$$

If a given $\lambda \in C$ and $x \neq 0$ in the space satisfy the equation, they are called an *eigenvalue* and corresponding *eigenvector*, respectively, of the operator U.

The finite dimensional theory of eigenvalues of Hermitian operators[1] can be applied (using the Transfer Principle) to the Hermitian operator T' on space E. Thus we may write the eigenvalues of T' as $\lambda_1, \ldots, \lambda_\nu$ (permitting repetitions) where $|\lambda_1| \geq |\lambda_2| \geq \cdots \geq |\lambda_\nu|$. There are (again using the Transfer Principle) corresponding eigenvectors $r_1 r_2, \ldots, r_\nu$ that are orthonormal (i.e., $(r_i,r_j) = 0$ for $i \neq j$, $\|r_i\| = 1$). And the eigenvalue equation holds: $(\lambda_i I - T')r_i = 0$, or equivalently, $T'r_i = \lambda_i r_i$, for $i = 1,2, \ldots, \nu$.

LEMMA 5. For $i \leq \nu$, $|\lambda_i| \leq \|T\|$. Hence each λ_i is finite.

[1] See Halmos [5], Sect. 79.

PROOF. Since $\|r_i\| = 1$,

$$\begin{aligned}|\lambda_i| &= |\lambda_i|\,\|r_i\| \\ &= \|\lambda_i r_i\| \\ &= \|T'r_i\| \\ &\le \|T'\| \cdot \|r_i\| \le \|T\|. \quad \blacksquare\end{aligned}$$

LEMMA 6. If $\lambda_j \not\approx 0$, then r_j is near standard.

PROOF. From the eigenvalue equation, $T'r_j = \lambda_j r_j$. By Lemma 4, $\lambda_j r_j$ is near standard, say $\lambda_j r_j \approx t_j \in H$. Let $\mu_j = {}^\circ(\lambda_j) \neq 0$. Then

$$r_j = \left(\frac{1}{\lambda_j}\right)\lambda_j r_j \approx \frac{1}{\mu_j} t_j \in H. \quad \blacksquare$$

LEMMA 7. Let ε be a positive real number, and let $|\lambda_j| \ge \varepsilon$ for all $j \le k$. Then $k \in N^+$.

PROOF. Since $|\lambda_j| \ge \varepsilon$, we have $\lambda_j \not\approx 0$, and hence by Lemma 6, r_j is near standard for all $j \le k$. By the Pythagorean theorem, $\|r_i - r_j\| = \sqrt{2}$ for $i,j \le k$. Suppose that $k \in {}^*N - N$. Then the internal sequence

$$s_i = \begin{cases} r_i & \text{for} \quad i \le k \\ 0 & \text{for} \quad i > k \end{cases}$$

has the property that $\|s_i - s_j\| \ge \sqrt{2}$ for $i \neq j$ and $i, j \in {}^*N$. By Theorem 3–5.21, some s_i, $i \le k$, is remote. But this contradicts the fact that r_i is near standard. $\blacksquare$

LEMMA 8. Let $\lambda_{r+1} = \lambda_{r+2} = \cdots = \lambda_{r+k} \not\approx 0$. Then $k \in N^+$.

PROOF.

$$|\lambda_1| \ge |\lambda_2| \ge \cdots \ge |\lambda_{r+1}| = |\lambda_{r+2}| = \cdots = |\lambda_{r+k}|,$$

and hence for all $j \le r + k$, $|\lambda_j| \ge \varepsilon$ for some positive real ε. By Lemma 7, $r + k \in N^+$, and hence $k \in N^+$. $\blacksquare$

Now we rewrite the sequence $\{\lambda_i | i \le \nu\}$ without repetitions as $\{\kappa_i | i \le \mu\}$. Thus each κ_i is equal to at least one λ_j,

$$|\kappa_1| \ge |\kappa_2| \ge \cdots \ge |\kappa_\mu|,$$

where $\mu \le \nu$, and each λ_i is equal to exactly one κ_j. Let the eigenvectors corresponding to κ_j be $r_1^{(j)}, r_2^{(j)}, \ldots, r_{m_j}^{(j)}$.

We can thus write

$$E_j = \operatorname{span}(r_1^{(j)}, \ldots, r_{m_j}^{(j)}),$$

so that by the Transfer Principle, E_j is a linear subspace of E (the *eigenspace* of κ_j). Using the spectral theorem for finite dimensional spaces (see e.g. Halmos [5], p. 156), recalling that T' is a Hermitian operator on E, and using the Transfer Principle, we have

$$E = E_1 \oplus E_2 \oplus \cdots \oplus E_\mu.$$

So by Corollary 2.4, we have

$$P = P_{E_1} + P_{E_2} + \cdots + P_{E_\mu}.$$

From this we can get the *spectral resolution* of T':

$$T' = \kappa_1 P_{E_1} + \kappa_2 P_{E_2} + \cdots + \kappa_\mu P_{E_\mu}.$$

To verify this equation note that $T'u = \kappa_j u$ for any $u \in E_j$. (This is because for $u = \alpha_1 r_1^{(j)} + \cdots + \alpha_{m_j} r_{m_j}^{(j)}$,

$$\begin{aligned} T'u &= \alpha_1 T' r_1^{(j)} + \cdots + \alpha_{m_j} T' r_{m_j}^{(j)} \\ &= \alpha_1 \kappa_j r_1^{(j)} + \cdots + \alpha_{m_j} \kappa_j r_{m_j}^{(j)} = \kappa_j u.) \end{aligned}$$

Thus

$$\begin{aligned} T'x = T'Px &= T'(P_{E_1}x + P_{E_2}x + \cdots + P_{E_\mu}x) \\ &= \kappa_1 P_{E_1}x + \kappa_2 P_{E_2}x + \cdots + \kappa_\mu P_{E_\mu}x. \end{aligned}$$

LEMMA 9. There is a $b \in E$ and a real $\varepsilon > 0$ such that

$$\|b\| \le 1 \quad \text{and} \quad \|T'b\| > \varepsilon.$$

PROOF. Since $T \neq 0$, there is a $c \in H$, $c \neq 0$ such that $Tc \neq 0$. Let $a = c/\|c\|$, $\alpha = \|Ta\|$. Thus $\alpha = \|Tc\|/\|c\| > 0$. Let $b = Pa \in E$. Since a is standard, $b \approx a$ by Lemma 4, and thus $Tb \approx Ta$. Hence

$$\|Tb\| \approx \|Ta\| = \alpha; \qquad \|Tb\| > \tfrac{1}{2}\alpha.$$

Then

$$\begin{aligned} \|T'b\| &= \|PTb\| \\ &\approx \|Tb\| > \tfrac{1}{2}\alpha, \end{aligned}$$

since Tb is near standard. So

$$\|T'b\| > \tfrac{1}{4}\alpha.$$

Finally,

$$\|b\| = \|Pa\| \le \|a\| = 1. \quad \blacksquare$$

We now define:

$$W = \{j \in {}^*N^+ | \kappa_j \approx 0\}.$$

Since this definition involves the external "$\approx$", W may well be external. Nevertheless, we shall see that W contains only finite numbers.

LEMMA 10. $W \neq \varnothing$.

PROOF. If $W = \varnothing$, then $\kappa_j \approx 0$ for all $j \in {}^*N^+$. Thus $\kappa = |\kappa_1| \approx 0$. Choose ε and b as in the previous lemma. By the above discussion we can write:

$$b = b_1 + b_2 + \cdots + b_\mu, \qquad b_j \in E_j,$$
$$T'b = \kappa_1 b_1 + \kappa_2 b_2 + \cdots + \kappa_\mu b_\mu.$$

So, using the Pythagorean theorem twice,

$$\begin{aligned} \|T'b\|^2 &= \sum_{i=1}^{\mu} |\kappa_i|^2 \|b_i\|^2 \\ &\leq \kappa^2 \sum_{i=1}^{\mu} \|b_i\|^2 \\ &= \kappa^2 \|b\|^2 \leq \kappa^2 \approx 0. \end{aligned}$$

This contradicts Lemma 9. ■

LEMMA 11. $W \subseteq N^+$.

PROOF. Let $m \in W$. Then $|\kappa_m| \geq \varepsilon$ for some real $\varepsilon > 0$. Then $|\kappa_j| > \varepsilon$ for all $j \leq m$. But by Lemma 7, this implies $m \in N^+$. ■

Now it is clear that if $m \in W$ and $1 \leq j \leq m$, then $j \in W$. So by Lemmas 10 and 11, there are two possibilities. Either

$$W = N^+ \quad \text{(and } W \text{ is external)},$$

or

$$W = \{1,2,\ldots,k\} \quad \text{for some} \quad k \in N \quad \text{(and } W \text{ is internal)}.$$

Our analysis splits into two cases, depending on whether W is finite or infinite.

LEMMA 12. If $x \in {}^*H$ is near standard, then $T'Px \approx Tx$.

PROOF. By Lemma 1, $Px \approx x$, so that

$$Tx \approx TPx.$$

Since Tx is near standard (e.g., by Theorem 4–1.10), so is TPx. Using Lemma 1 again, $TPx \approx P(TPx)$. Thus

$$Tx \approx PTPx = T'Px. \quad ■$$

LEMMA 13. If $u,v \in {}^*H$ are near standard, then ${}^\circ(u,v) = ({}^\circ u, {}^\circ v)$.

PROOF. This is just the nonstandard formulation of the continuity of the inner product.

THEOREM 4.1. If $\lambda_j \not\approx 0$, then λ_j is finite, r_j is near standard, and each $°(\lambda_j)$, $°(r_j)$ are an eigenvalue and the corresponding eigenvector of T in H.

PROOF. By Lemmas 5 and 6, λ_j is finite and r_j is near standard. Using Lemma 12, and the fact that $Pr_j = r_j$, we have

$$\begin{aligned} T°r_j &\approx Tr_j \\ &\approx T'Pr_j = T'r_j \\ &= \lambda_j r_j \\ &\approx °\lambda_j \cdot °r_j. \end{aligned}$$

Finally, $°r_j \neq 0$, since by Lemma 13,

$$(°r_j,°r_j) = °(r_j,r_j) = 1. \quad \blacksquare$$

Note that if $i \neq j$ and $\lambda_i,\lambda_j \not\approx 0$ then

$$(°r_i,°r_j) = °(r_i,r_j) = \delta^i_j \qquad (\ddagger)$$

Now, by Lemmas 6 and 8, for $j \in W$, the vectors $r_1^{(j)}, \ldots, r_{m_j}^{(j)}$ are all near standard and m_j is finite. Let $s_i^{(j)} = °(r_i^{(j)})$, $i = 1,2, \ldots, m_j$, and let

$$\tilde{H}_j = \operatorname{span}(s_1^{(j)}, \ldots, s_{m_j}^{(j)}) \subset H.$$

Then, by ($\ddagger$), $\{s_i^{(j)} | 1 \leq i \leq m_j\}$ is an orthonormal sequence and hence linearly independent. Thus $\tilde{H}_j$ has dimension m_j. Also, using ($\ddagger$), we have that for $j,k \in W, j \neq k$,

$$\tilde{H}_j \perp \tilde{H}_k.$$

Moreover, by Lemma 15 of Section 2, we have

$$P_{\tilde{H}_j}x = \sum_{i=1}^{m_j} (x,s_i^{(j)})s_i^{(j)}.$$

LEMMA 14. For $x \in H, j \in W, P_{\tilde{H}_j}x \approx P_{E_j}x$.

PROOF. Using Lemma 13, Lemma 15 of Section 2, and the Transfer Principle,

$$P_{\tilde{H}_j}x = \sum_{i=1}^{m_j} (x,s_i^{(j)})s_i^{(j)} \approx \sum_{i=1}^{m_j} (x,r_i^{(j)})r_i^{(j)} = P_{E_j}x. \quad \blacksquare$$

Now let

$$K = \tilde{H}_1 \oplus \tilde{H}_2 \oplus \cdots,$$

where this direct sum is finite or infinite according to whether W is, and let $\tilde{H}_0 = K^\perp$. Then by the discussion in Section 2 (Corollary 2.4 or Lemma 14, respectively), we have

$$H = \tilde{H}_0 \oplus \tilde{H}_1 \oplus \tilde{H}_2 \oplus \cdots,$$

and for each $x \in H$,

$$x = P_{\tilde{H}_0}x + P_{\tilde{H}_1}x + P_{\tilde{H}_2}x + \cdots.$$

For each $j \in W$, let us write $v_j = {}^\circ(\kappa_j)$. Then we have

THEOREM 4.2. [SPECTRAL THEOREM FOR COMPACT HERMITIAN OPERATORS (PART I)]. If $W = \{1,2,\ldots,k\}$, then the eigenvalues of T are $0,v_1,\ldots,v_k$, and

$$T = v_1 P_{\tilde{H}_1} + v_2 P_{\tilde{H}_2} + \cdots + v_k P_{\tilde{H}_k}.$$

PROOF. We begin by calculating Tx for $x \in H$. By Lemma 2, $Tx = T'x$. Hence using the spectral resolution of T',

$$\begin{aligned} Tx &= \sum_{j \le \mu} \kappa_j P_{E_j} x \\ &= \sum_{j=1}^{k} \kappa_j P_{E_j} x + \sum_{j=k+1}^{\mu} \kappa_j P_{E_j} x \\ &= I + II. \end{aligned}$$

Using Lemma 14,

$$I \approx \sum_{j=1}^{k} v_j P_{\tilde{H}_j} x.$$

So we need to show that $II \approx 0$. But, using the Pythagorean theorem twice,

$$\begin{aligned} \|II\|^2 &= \left\| \sum_{j=k+1}^{\mu} \kappa_j P_{E_j} x \right\|^2 \\ &= \sum_{j=k+1}^{\mu} |\kappa_j|^2 \|P_{E_j} x\|^2 \\ &\le |\kappa_{k+1}|^2 \sum_{j=1}^{\mu} \|P_{E_j} x\|^2 \\ &= |\kappa_{k+1}|^2 \|x\|^2 \approx 0, \end{aligned}$$

since

$$x = \sum_{j=1}^{\mu} P_{E_j} x.$$

Next note that 0 is an eigenvector of T, since for any $x \in \tilde{H}_0$,

$$Tx = \sum_{j=1}^{k} v_j P_{\tilde{H}_j} x = 0.$$

Finally, let λ be any eigenvalue of T, r a corresponding eigenvector. It remains to show that if $\lambda \neq 0$, then $\lambda = \nu_j$ for some $j = 1, \ldots, k$. Now,

$$\lambda r = \lambda \sum_{j=0}^{k} P_{\tilde{H}_j} r,$$

and

$$\lambda r = Tr = \sum_{j=1}^{k} \nu_j P_{H_j} r.$$

By the uniqueness of the direct sum decomposition, we have

$$\lambda P_{\tilde{H}_0} r = 0; \qquad \lambda P_{\tilde{H}_j} r = \nu_j P_{\tilde{H}_j} r, j > 0.$$

If $\lambda \neq 0$, then $P_{H_0} r = 0$, so that for some (unique) $j > 0$, $P_{H_j} r \neq 0$ and $\lambda = \nu_j$. ■

If W is infinite we cannot say quite so much:

THEOREM 4.3. SPECTRAL THEOREM FOR COMPACT HERMITIAN OPERATORS (PART II). If $W = N$, then for all $x \in H$,

$$Tx = \sum_{i=1}^{\infty} \nu_i P_{\tilde{H}_i} x.$$

We first obtain

LEMMA 15. If $W = N^+$, then $\lim_{j \to \infty} \nu_j = 0$.

PROOF. $|\nu_j| \approx |\kappa_j|$ for all $j \in N^+$. Hence by the Infinitesimal Prolongation Theorem, $|\nu_n| \approx |\kappa_n| \approx 0$ for some infinite n. So, 0 is a limit point of $\{|\nu_j| \mid j \in N^+\}$. On the other hand, this sequence is monotone decreasing (since $\{|\kappa_j| \mid j \in N^+\}$ is, and $\nu_j = {}^\circ\kappa_j$) and hence convergent. Therefore, $|\nu_j| \to 0$. ■

PROOF OF THEOREM. Let $x \in H$, and set

$$s_n = \left\| Tx - \sum_{j=1}^{n} \nu_j P_{\tilde{H}_j} x \right\|.$$

We wish to show that $s_n \to 0$.

Now, by Lemma 14, for n finite,

$$\sum_{j=1}^{n} \nu_i P_{H_i} x \approx \sum_{j=1}^{n} \kappa_j P_{E_j} x.$$

Also,

$$Tx = T'x = \sum_{j=1}^{\mu} \kappa_j P_{E_j} x.$$

So,

$$s_n \approx \left\| \sum_{j=n+1}^{\mu} \kappa_j P_{E_j} x \right\| \quad \text{for all finite } n.$$

But using the Pythagorean theorem,

$$\begin{aligned} \left\| \sum_{j=n+1}^{\mu} \kappa_j P_{E_j} x \right\|^2 &= \sum_{j=n+1}^{\mu} |\kappa_j|^2 \|P_{E_j} x\|^2 \\ &\leq |\kappa_{n+1}|^2 \sum_{j=1}^{\mu} \|P_{E_j} x\|^2 \\ &= |\kappa_{n+1}|^2 \|x\|^2. \end{aligned}$$

So for all finite n,

$$s_n \leq |\nu_{n+1}| \cdot \|x\|,$$

and by Lemma 15, $s_n \to 0$. ■

5. NONCOMPACT HERMITIAN OPERATORS

Now we obtain the spectral theorem for not necessarily compact Hermitian operators on a Hilbert space H. (We are following Bernstein [2].) Once again, it is a question of associating with such an operator a family of projections in terms of which the operator can be expressed. However, in the general case, the operator is "resolved" by an integral, rather than a sum.

To begin with, it is necessary to explain the kind of integral used. Thus let $I = [a,b]$ be an interval $a < b$, $a,b \in R$. Let g map I into R, and let $V(\lambda)$ be a function defined on I whose values are bounded linear operators on H. Then we write:

$$T = \int_a^b g(\lambda)\, dV(\lambda)$$

to mean that for every real $\varepsilon > 0$, there is a corresponding real δ such that the inequality

$$\left\| T - \sum_{j=0}^{n-1} g(r_j)[V(t_{j+1}) - V(t_j)] \right\| < \varepsilon$$

holds whenever the following conditions hold:

(a) $a = t_0 < t_1 < \cdots < t_n = b$,
(b) $t_{i+1} - t_i < \delta$, $i = 0,1, \ldots ,n - 1$,
(c) $t_i \leq r_i \leq t_{i+1}$, $i = 0,1, \ldots ,n - 1$.

With this explained, we can state:

SPECTRAL THEOREM. Let T be a Hermitian operator on H. Let

$$a = \inf_{||x||=1} (Tx,x), \qquad b = \sup_{||x||=1} (Tx,x).$$

Then there is a family $\{E(\lambda)|\lambda \in R\}$ of projections on H such that

(1) $E(\lambda)E(\mu) = E(\lambda)$ if $\lambda \leq \mu$,
(2) $T \cdot E(\lambda) = E(\lambda) \cdot T$ for each $\lambda \in R$.
(3) Let $\{\kappa_n|n = 1,2,3, \ldots\}$ be a sequence of real numbers such that $\kappa_n > \lambda$ for each n and $\kappa_n \to \lambda$.

Then for every $x \in H$, $E(\kappa_n)x \to E(\lambda)x$,

(4) $E(\lambda) = 0$ for $\lambda < a$ and $E(\lambda) = I$ for $\lambda \geq b$.
(5) For each real $h > 0$,

$$T = \int_{a-h}^{b} \lambda \, dE(\lambda).$$

Thus, below T is an arbitrary Hermitian operator on H. E,P, and T' are defined as in the previous section. Once again, the finite dimensional spectral theorem can be applied to T' as a Hermitian operator on the space E. And the proofs of Lemmas 1, 2, 3, and 5 (but *not* of Lemma 4) of Section 4 hold without change.

Let us write J for the interval $[-||T||,||T||]$. Then by Lemma 5 of Section 4, for each i, $1 \leq i \leq \nu$, we have $\lambda_i \in {}^*J$. We also have

LEMMA 1. $a,b \in J$.

PROOF. For $||x|| = 1$,

$$|(Tx,x)| \leq ||Tx|| \cdot ||x|| = ||Tx||.$$

Hence $|b| = \left| \sup_{||x||=1} (Tx,x) \right| \leq \sup_{||x||=1} |(Tx,x)| \leq \sup_{||x||=1} ||Tx|| \cdot ||x|| = ||T||$. Also

$$\begin{aligned} a = \left| \inf_{||x||=1} (Tx,x) \right| &= \left| \sup_{||x||=1} (-Tx,x) \right| \\ &\leq \sup_{||x||=1} |(-Tx,x)| \leq \sup_{||x||=1} ||-Tx|| \cdot ||x|| \\ &= ||T||. \quad \blacksquare \end{aligned}$$

LEMMA 2. For $n \geq 0$, $(T')^n = \sum_{i=1}^{\nu} \lambda_i^n P_i$.

PROOF. The proof is by induction on n. The case $n = 0$ is trivial. Assuming the result holds for n, we obtain it for $n + 1$:

$$\begin{aligned}(T')^{n+1} &= (T')^n T' \\ &= \left(\sum_{i=1}^{\nu} \lambda_i^n P_i\right) \sum_{i=1}^{\nu} \lambda_i P_i \\ &= \sum_{i=1}^{\nu} \sum_{j=1}^{\nu} \lambda_i^n \lambda_j P_i P_j \\ &= \sum_{i=1}^{\nu} \lambda_i^{n+1} P_i,\end{aligned}$$

since $P_i P_j = 0$ for $i \neq j$ and $P_i^2 = P_i$. ■

For any polynomial $p(\lambda) = a_0 + a_1\lambda + \cdots + a_n\lambda^n$, $a_i \in {}^*R$, it follows from Lemma 2 that

$$p(T') = \sum_{i=1}^{\nu} p(\lambda_i) P_i.$$

This suggests the

DEFINITION. Let f be an internal map into ${}^*R$, whose domain includes ${}^*J$. Then we define the operator

$$f(T') = \sum_{i=1}^{\nu} f(\lambda_i) P_i,$$

LEMMA 3. Let f be as in the definition just stated. Then $f(T')$ is Hermitian.

PROOF. Using the fact that the P_is are Hermitian,

$$\begin{aligned}(f(T')x,y) &= \left(\sum_{i=1}^{\nu} f(\lambda_i) P_i x, y\right) \\ &= \sum_{i=1}^{\nu} f(\lambda_i)(P_i x, y) \\ &= \sum_{i=1}^{\nu} f(\lambda_i)(x, P_i y) \\ &= \left(x, \sum_{i=1}^{\nu} f(\lambda_i) P y\right) \\ &= (x, f(T')y),\end{aligned}$$

since each $f(\lambda_i) \in {}^*R$. ■

COROLLARY 5.2. Let f be a continuous map of J into R with $f(\lambda) \geq 0$ for all $\lambda \in J$; then ${}^\circ(f(T')) \geq 0$.

PROOF. By the Transfer Principle, $f(\lambda) \geq 0$ for $\lambda \in {}^*J$. By Lemma 5, $f(T') \geq 0$. Thus for $x \in E$, $(f(T')x,x) \geq 0$. Since $H \subseteq E$, this last inequality holds for all $x \in H$. The result follows by taking standard parts. ■

THEOREM 5.3. Let $\{f_n | n \in N^+\}$ be a sequence of continuous functions from J into R such that

(1) $f_n(\lambda) \geq 0$ for $\lambda \in J$,
(2) $f_{n+1}(\lambda) \leq f_n(\lambda)$, $\lambda \in J$.

Then for each $x \in H$, $\{{}^\circ(f_n(T'))x | n \in N^+\}$ has a limit.

PROOF. Set $F_n = {}^\circ(f_n(T'))$.

CLAIM 1.

For all $m,n \in N$,

$$F_n F_m = F_m F_n.$$

By the operational calculus,

$$f_n(T')f_m(T') = f_m(T')f_n(T').$$

The claim follows by taking standard parts.

CLAIM 2.

For $m \geq n$ and $\lambda \in J$,

$$f_n(\lambda)f_n(\lambda) \geq f_n(\lambda)f_m(\lambda) \geq f_m(\lambda)f_m(\lambda) \geq 0.$$

Since $f_n(\lambda) \geq f_m(\lambda) \geq 0$ and $f_n(\lambda) \geq 0$, the first inequality follows. The second inequality follows similarly.

By Corollary 5.2 and Claim 2, when $m \geq n$,

$$F_n^2 \geq F_n F_m \geq F_m^2 > 0. \tag{1}$$

Hence for $m \geq n$ and $x \in H$,

$$(F_n^2 x,x) \geq (F_m^2 x,x) \geq 0.$$

Therefore, the sequence

$$\{(F_n^2 x,x) | n \in N^+\}$$

is a monotone decreasing sequence of real numbers and hence converges. Thus it must be a Cauchy sequence. From (1), we have that for $m \geq n$ and $x \in H$,

$$(F_n^2 x,x) \geq (F_n F_m x,x) \geq (F_m^2 x,x) \geq 0. \tag{2}$$

Since the sequence $\{(F_n^2x,x)\}$ is a Cauchy sequence, for all $n,m \in {}^*N - N$,

$$(F_n^2x,x) \approx (F_m^2x,x)$$

and by (2)

$$(F_n^2x,x) \approx (F_nF_mx,x) \approx (F_m^2x,x). \tag{3}$$

Since the F_n are Hermitian, using Claim 1,

$$\begin{aligned}\|F_nx - F_mx\|^2 &= (F_nx,F_nx) - (F_nx,F_mx) - (F_mx,F_nx) + (F_mx,F_mx)\\ &= (F_n^2x,x) - 2(F_nF_mx,x) + (F_m^2x,x).\end{aligned}$$

Using (3), we see that for $n,m \in {}^*N - N$, $F_nx \approx F_mx$. Thus the sequence $\{F_nx\}$ is a Cauchy sequence and so has a limit. ■

For each $\eta \in {}^*R$ we set

$$S(\eta) = \sum_{\lambda_i < \eta} P_i.$$

By Corollary 2.4 and the Transfer Principle, each $S(\eta)$ is a projection on E. In order to deal with $S(\eta)$ via the operational calculus, we introduce the Heaviside unit function:

$$h_\eta(\lambda) = \begin{cases} 1 & \text{for} \quad \lambda < \eta \\ 0 & \text{for} \quad \lambda \geq \eta. \end{cases}$$

LEMMA 12. $S(\eta) = h_\eta(T')$.

PROOF.

$$h_\eta(T') = \sum_{i=1}^{\nu} h_\eta(\lambda_i)P_i = \sum_{\lambda_i < \eta} P_i = S(\eta). \quad ■$$

LEMMA 13. Let $\eta_1 < \eta_2 < \eta_3$, and let $x \in E$. Then

$$\|S(\eta_2)x - S(\eta_1)x\| \leq \|S(\eta_3)x - S(\eta_1)x\|.$$

PROOF. The given inequality implies that for all λ,

$$h_{\eta_1}(\lambda) \leq h_{\eta_2}(\lambda) \leq h_{\eta_3}(\lambda).$$

The result then follows from Lemmas 6 and 12. ■

THEOREM 5.4. Let x be in H and $\mu \in R$. Then there is an $\eta > 0$, $\eta \approx 0$, $\eta \in {}^*R$ such that

(1) $S(\mu + \eta)x$ is near standard,
(2) $\eta' > \eta$, $\eta' \approx 0$ implies $S(\mu + \eta)x \approx S(\mu + \eta')x$.

(Note that, in general, η depends on both μ and x.)

PROOF. We define

$$g_n(\lambda) = \begin{cases} 1 & \text{for} \quad \lambda \le \mu + [1/(n-1)] \\ n(n-1)[\lambda - (\mu + (1/n))] & \text{for} \quad \mu + [1/(n-1)] < \lambda \le \mu + (1/n) \\ 0 & \text{for} \quad \lambda > \mu + (1/n). \end{cases}$$

That is, $g_n(\lambda)$ is 1 for $\lambda \le \mu + [1/(n-1)]$, is 0 for $\lambda \ge \mu + (1/n)$, and drops linearly from 1 to 0 on $[\mu + 1/(n-1), \mu + (1/n)]$. Thus each $g_n(\lambda)$ is continuous on R. Moreover (the reader may want to draw a diagram), $g_{n+1}(\lambda) \le g_n(\lambda)$ for all $\lambda \in R$. Hence by previous results we have

(1) $g_n(T')x$ is near standard for $n = 1,2,3, \ldots$.
(2) The sequence $\{y_n | n \in N^+\}$, where $y_n = {}^\circ g_n(T')x$, converges to some $y \in H$.

Then $y_n \approx y$ for all $y \in {}^*N - N$. Since the sequence $\{g_n(T')x\}$ is internal and $y_n \approx g_n(T')x$ for all finite n, by the Infinitesimal Prolongation Theorem there exists a $\kappa \in {}^*N - N$, such that $y_n \approx g_n(T')x$ for $n \le \kappa$. Thus $y \approx g_n(T')x$ for infinite $n \le \kappa$. From this discussion, we conclude that

(a) $g_n(T')x$ is near standard for *all* $n \le \kappa$,
(b) $g_n(T')x \approx g_m(T')x$ for infinite $m,n \le \kappa$.

It is easily verified (draw a diagram!) that for $\lambda \in R$ and $n \in N^+$,

$$g_n(\lambda) \le h_{\mu+(1/n)}(\lambda) \le g_{n-1}(\lambda).$$

Now set $\eta = 1/\kappa$, so $\eta \approx 0$. Using the Transfer Principle,

$$g_\kappa(\lambda) \le h_{\mu+\eta}(\lambda) \le g_{\kappa-1}(\lambda),$$

for all $\lambda \in {}^*R$. Using Lemma 6,

$$\|h_{\mu+\eta}(T')x - g_\kappa(T')x\| \le \|g_{\kappa-1}(T')x - g_\kappa(T')x\|.$$

By fact (b), the right side is infinitesimal, and hence so is the left side. Therefore,

$$S(\mu + \eta)x = h_{\kappa+\eta}(T')x \approx g_\kappa(T')x, \tag{4}$$

and by fact (a), $S(\mu + \eta)$ is near standard.

To complete the proof of the theorem, choose $\eta' > \eta$ where $\eta' \approx 0$. Since $1/\eta'$ is infinite, there exists (by the Transfer Principle) an $m \in {}^*N - N$ so that $m - 1 < 1/\eta' \le m$, and hence $1/m \le \eta' < 1/(m-1)$. Since $\eta' > \eta$, $1/\eta' < 1/\eta = \kappa$, and hence $m - 1 < \kappa$; so $m \le \kappa$.

Now, it is easy to check that (please draw one last diagram!) if $n \in N$ $(n > 2)$ and $1/n < t < 1/(n-1)$, then

$$g_n(\lambda) \le h_{\mu+t}(\lambda) \le g_{n-2}(\lambda).$$

Using the Transfer Principle and Lemma 6, we have

$$\|h_{\mu+\eta'}(T')x - g_m(T')x\| \leq \|g_{m-2}(T')x - g_m(T')x\|.$$

Since $m - 2 < m \leq \kappa$, by fact (b) the right side, and thus the left side, is infinitesimal. Therefore,

$$S(\mu + \eta')x = h_{\mu+\eta'}(T')x \approx g_m(T')x \approx g_\kappa(T')x \approx S(\mu + \eta)x,$$

using fact (b) and (4).

This completes the proof. ∎

As a result of the previous theorem, given any $\lambda \in R$ and $x \in H$, there corresponds an infinitesimal $\eta_x^\lambda > 0$ such that

(1) *$s(\lambda + \eta_x^\lambda)x$ is near standard,*
(2) *$\eta > \eta_x^\lambda$, $\eta \approx 0$ implies $S(\lambda + \eta)x \approx S(\lambda + \eta_x^\lambda)x$.*

We omit the superscript λ when there is no ambiguity.

Now we are ready to define the family of (as it will turn out) projections on H, $\{E(\lambda)|\lambda \in R\}$ that occurs in the statement of the spectral theorem. Namely, we set

$$E(\lambda)x = {}^\circ(S(\lambda + \eta_x^\lambda)x).$$

LEMMA 14. $E(\lambda)$ is a bounded linear operator on H.

PROOF. Let $x,y \in H$. Then we have with $\eta \approx 0$,

$$\begin{aligned} E(\lambda)x &\approx S(\lambda + \eta)x \quad &\text{for} \quad \eta > \eta_x, \\ E(\lambda)y &\approx S(\lambda + \eta)y \quad &\text{for} \quad \eta > \eta_y, \\ E(\lambda)(x + y) &\approx S(\lambda + \eta)(x + y) \quad &\text{for} \quad \eta > \eta_{x+y}. \end{aligned}$$

Hence all three relations hold for any infinitesimal η, which is $> \eta_x, \eta_y, \eta_{x+y}$. Since $S(\lambda + \eta)$ is linear, we have, using such an η,

$$E(\lambda)(x + y) \approx E(\lambda)x + E(\lambda)y,$$

and the "$\approx$" can be replaced by "$=$", since both sides are standard.

Similarly, for $x \in H$, $\alpha \in C$, we can write:

$$\begin{aligned} E(\lambda)(\alpha x) &\approx S(\lambda + \eta)(\alpha x) \\ &= \alpha S(\lambda + \eta)x \quad \text{for} \quad \eta > \eta_x, \end{aligned}$$

and

$$\alpha E(\lambda)x \approx \alpha S(\lambda + \eta)x \quad \text{for} \quad \eta > \eta_x.$$

Hence

$$E(\lambda)(\alpha x) = \alpha E(\lambda)x.$$

Finally,

$$\|S(\lambda + \eta_x)x\| \leq \|x\|$$

since $\|S(\mu)\| = 1$ for all $\mu \in {}^*R$. Taking standard parts,

$$\|E(\lambda)x\| \leq \|x\|. \quad \blacksquare$$

LEMMA 15. $E^2(\lambda) = E(\lambda)$.

PROOF. Let $y = E(\lambda)x$. Choose $\eta > \max\{\eta_x, \eta_y\}$ with $\eta \approx 0$. Then

$$\begin{aligned} E^2(\lambda)x &= E(\lambda)y \\ &\approx S(\lambda + \eta)y \\ &= S(\lambda + \eta)(E(\lambda)x) \\ &\approx S(\lambda + \eta)S(\lambda + \eta)x \\ &= S(\lambda + \eta)x \\ &\approx E(\lambda)x, \end{aligned}$$

where we have used the fact that $S^2(\lambda + \eta) = S(\lambda + \eta)$, since this operator is a projection. ■

THEOREM 5.5. For each $\lambda \in R$, $E(\lambda)$ is a projection on H.

PROOF. By Theorem 2.5 and the previous two lemmas, we need only verify that $E(\lambda)$ is Hermitian. But this follows easily by a calculation using the continuity of the inner product. Namely, let $x,y \in H$, and let $\eta \approx 0$, $\eta > \eta x, \eta_y$. Then

$$\begin{aligned} (E(\lambda)x,y) &\approx (S(\lambda + \eta)x,y) \\ &\approx (x,S(\lambda + \eta)y) \\ &\approx (x,E(\lambda)y). \quad \blacksquare \end{aligned}$$

LEMMA 16. For $\eta_1 < \eta_2$,

$$S(\eta_1) \cdot S(\eta_2) = S(\eta_1).$$

PROOF. For $\eta_1,\eta_2 \in R$, $\eta_1 < \eta_2$, we have, obviously,

$$h_{\eta_1}(\lambda) \cdot h_{\eta_2}(\lambda) = h_{\eta_1}(\lambda)$$

for $\lambda \in R$. The result follows by the Transfer Principle and the operational calculus. ■

LEMMA 17. For $\mu \leq \kappa$,

$$E(\mu) \cdot E(\kappa) = E(\mu).$$

PROOF. For the case $\mu = \kappa$, the result follows because $E(\mu)$ is idempotent. Let $\mu < \kappa$. Choose $\eta > \eta_x^\mu, \eta_x^\kappa, \eta_{E(\kappa)x}^\mu$ with $\eta \approx 0$. Then

(1) $E(\mu)x \approx S(\mu + \eta)x$,
(2) $E(\kappa)x \approx S(\kappa + \eta)x$.

Now,

$$\begin{aligned} E(\mu)E(\kappa)x &\approx S(\mu + \eta)E(\kappa)x \\ &\approx S(\mu + \eta)S(\kappa + \eta)x \\ &= S(\mu + \eta)x \\ &\approx E(\mu)x, \end{aligned}$$

where we have used Lemma 16. ■

LEMMA 18. $T \cdot E(\lambda) = E(\lambda) \cdot T$.

PROOF. We first note that by the operational calculus,

$$S(\zeta) \cdot T' = h_\zeta(T') \cdot T' = T' \cdot h_\zeta(T') = T' \cdot S(\zeta),$$

for all $\zeta \in {}^*R$.

Now let $x \in H$. Let $\eta \approx 0, \eta > \eta_x, \eta_{Tx}$. Then

$$\begin{aligned} (T \cdot E(\lambda))x &= T'(E(\lambda)x) \\ &\approx T'(S(\lambda + \eta)x) \\ &= S(\lambda + \eta)T'x \\ &= S(\lambda + \eta)Tx \\ &\approx E(\lambda)(Tx) \\ &= (E(\lambda) \cdot T)x. \end{aligned}$$ ■

LEMMA 19. Let κ_n be a *strictly decreasing* sequence such that $\kappa_n \to \lambda$. Then $E(\kappa_n)x \to E(\lambda)x$ for each $x \in H$.

PROOF. By Lemma 17 and Theorem 2.6, $E(\kappa_{n+1}) \leq E(\kappa_n)$. Hence by Lemma 14 of Section 2, the sequence $\{E(\kappa_n)x | n \in N^+\}$ converges for each $x \in H$. Hence it suffices to show that for each $x \in H$, we have $E(\kappa_\nu)x \approx E(\lambda)x$ for *some* infinite integer ν.

Thus we choose positive infinitesimals η_n, $n = 1,2,3, \ldots$, and η, for which

$$E(\kappa_n)x \approx S(\kappa + \eta_n)x, \qquad E(\lambda)x \approx S(\lambda + \eta)x,$$

where the second relation holds as well for any positive infinitesimal $>\eta$. Then since $\lambda < \kappa_n$ for each n, we have

$$\lambda + \eta < \kappa_n < \kappa_n + \eta_n < \kappa_n + \frac{1}{n}$$

For each $n \in N^+$, we have

$$\left\| S\left(\kappa_n + \frac{1}{n}\right)x - E(\lambda)x \right\| \leq \left\| S\left(\kappa_n + \frac{1}{n}\right)x - S(\lambda + \eta)x \right\| + \|S(\lambda + \eta)x - E(\lambda)x\|.$$

Therefore,

$$\left\| S\left(\kappa_n + \frac{1}{n}\right)x - E(\lambda)x \right\| = \left\| S\left(\kappa_n + \frac{1}{n}\right)x - S(\lambda + \eta)x \right\| + \text{infinitesimal.}$$

Similarly,

$$\left\| S\left(\kappa_n + \frac{1}{n}\right)x - E(\kappa_n)x \right\| = \left\| S\left(\kappa_n + \frac{1}{n}\right)x - S(\kappa_n + \eta_n)x \right\| + \text{infinitesimal.}$$

Now by Lemma 13,

$$\left\| S\left(\kappa_n + \frac{1}{n}\right)x - S(\lambda + \eta)x \right\| - \left\| S\left(\kappa_n + \frac{1}{n}\right)x - S(\kappa_n + \eta_n)x \right\| \geq 0.$$

Hence the quantity

$$c_n = \left\| S\left(\kappa_n + \frac{1}{n}\right)x - E(\lambda)x \right\| - \left\| S\left(\kappa_n + \frac{1}{n}\right)x - E(\kappa_n)x \right\| \tag{5}$$

is either ≥ 0 or infinitesimal for all $n \in N^+$. Let c_n be defined by (5) for all $n \in {}^*N^+$, and let $d_n = |c_n| - c_n$. Thus if $c_n \geq 0$, then $d_n = 0$, and if $c_n \approx 0$, then $d_n \approx 0$. Hence $d_n \approx 0$ for $n \in N^+$, and so by the Infinitesimal Prolongation Theorem, $d_n \approx 0$ for all $n \leq \mu$ where $\mu \in {}^*N - N$. Choose $\nu \in {}^*N - N$ so $\nu < \mu$ and $\nu < 1/\eta$. This can be done because μ and $1/\eta$ are both positive infinite hyperreal numbers. Thus $1/\nu < \eta$, so that $E(\lambda)x \approx S(\lambda + \eta')x$ if $\eta' > 1/\nu$. Also, since $\kappa_n \to \lambda$ and $\kappa_n > \lambda$, we have

$$(\kappa_\nu - \lambda)\begin{cases} \approx 0 \\ > 0. \end{cases}$$

Hence

$$(\kappa_\nu - \lambda + 1/\nu)\begin{cases} \approx 0 \\ > 1/\nu. \end{cases}$$

Thus

$$\begin{aligned} E(\lambda)x &\approx S\left(\lambda + \left(\kappa_\nu - \lambda + \frac{1}{\nu}\right)\right)x \\ &= S\left(\kappa_\nu + \frac{1}{\nu}\right)x. \end{aligned}$$

Hence by (5)

$$\left\| S\left(\kappa_\nu + \frac{1}{\nu}\right)x - E(\kappa_\nu)x \right\| = \delta - c_\nu,$$

where $\delta \approx 0$. Since $\delta - c_\nu \geq 0$, c_ν must be infinitesimal. (If $c_\nu < 0$ and

$c_\nu \not\approx 0$, then $d_\nu = 2c_\nu \not\approx 0$.) So, $\delta - c_\nu \approx 0$ and

$$S\left(\kappa_\nu + \frac{1}{\nu}\right)x \approx E(\kappa_\nu)x.$$

Hence finally,

$$E(\kappa_\nu)x \approx E(\lambda)x. \quad \blacksquare$$

LEMMA 20. Let $\lambda_1 < \lambda_2 < \lambda_3$. Then

$$\|E(\lambda_2)x - E(\lambda_1)x\| \leq \|E(\lambda_3)x - E(\lambda_1)x\|.$$

PROOF. We write:

$$E(\lambda_1) \approx S(\lambda_i + \eta)x, \qquad i = 1,2,3,$$

where $\eta > 0$, $\eta \approx 0$. By Lemma 13,

$$\|S(\lambda_2 + \eta)x - S(\lambda_1 + \eta)x\| \leq \|S(\lambda_3 + \eta)x - S(\lambda_1 + \eta)x\|.$$

Taking standard parts, we obtain the result. $\blacksquare$

LEMMA 21. Let $\kappa_n \to \lambda$ where $\kappa_n > \lambda$ for all n. Then $E(\kappa_n)x \to E(\lambda)x$ for $x \in H$.

PROOF. Let $\{\kappa_{n_i} | i \in N^+\}$ be some strictly decreasing subsequence of $\{\kappa_n | n \in N^+\}$. (For example, assuming that $\kappa_{n_1}, \ldots, \kappa_{n_k}$ have been defined, n_{k+1} can be chosen to be the least subscript $> n_k$ such that $\kappa_{n_{k+1}} < \kappa_{n_k}$. The hypothesis guarantees that such an n_{k+1} always exists.) By Lemma 19, $E(\kappa_{n_i})x \to E(\lambda)x$. Choose any real $\varepsilon > 0$. Then for some i,

$$\|E(\kappa_{n_i})x - E(\lambda)x\| < \varepsilon.$$

Let n_0 be such that $\kappa_n < \kappa_{n_i}$ for all $n > n_0$. Thus by the previous lemma, $n > n_0$ implies:

$$\|E(\kappa_n)x - E(\lambda)x\| \leq \|E(\kappa_{n_i})x - E(\lambda)x\| < \varepsilon,$$

That is, $E(\kappa_n)x \to E(\lambda)x$. $\blacksquare$

We have just proved (3) of the Spectral Theorem stated at the beginning of this section. (1) was Lemma 17, and (2) was Lemma 18. The statements of (4) and (5) involve the numbers a,b defined in the statement of the theorem. The proof of (4) is contained in the following lemmas:

LEMMA 22. If $\lambda > b$, then $E(\lambda) = I$.

PROOF. Suppose $\lambda > b$, but that $E(\lambda) \neq I$. Since $E(\lambda)$ is a projection, $E(\lambda) = P_F$ for some closed linear subspace F of H. Thus (cf. Corollary 2.3) $I - E(\lambda) = P_{F^\perp} \neq 0$, and hence $F^\perp \neq \{0\}$. So let $x \in F^\perp$ with $\|x\| = 1$. Since $x \in F^\perp$,

$$(I - E(\lambda))x = x \quad \text{and} \quad E(\lambda)x = 0.$$

We have

$$E(\lambda)x \approx S(\lambda + \eta)x$$

for some positive $\eta \approx 0$. Let $W = \{i|\lambda + \eta \le \lambda_i\}$. Then

$$\begin{aligned} x &= x - E(\lambda)x \\ &\approx \sum_{i=1}^{v} P_i x - S(\lambda + \eta)x \\ &= \sum_{i=1}^{v} P_i x - \sum_{\lambda_i < \lambda + \eta} P_i x \\ &= \sum_{i \in W} P_i x. \end{aligned}$$

Using the spectral resolution of T' and the fact that x is standard, we have

$$\begin{aligned} Tx &= T'x \\ &\approx T'\left(\sum_{i \in W} P_i x\right) \\ &= \sum_{j=1}^{v} \lambda_j P_j \left(\sum_{i \in W} P_i x\right) \\ &= \sum_{j \in W} \lambda_j P_j x. \end{aligned}$$

Recalling the definition of W and using the continuity of the inner product, we have

$$\begin{aligned} (Tx,x) &\approx \left(\sum_{i \in W} \lambda_i P_i x, \sum_{i \in W} P_i x\right) \\ &= \sum_{i \in W} \lambda_i (P_i x, P_i x) \\ &= \sum_{i \in W} \lambda_i \|P_i x\|^2 \\ &\ge (\lambda + \eta) \sum_{i \in W} \|P_i x\|^2 \\ &= (\lambda + \eta) \left\|\sum_{i \in W} P_i x\right\|^2 \\ &\approx (\lambda + \eta)\|x\|^2 \\ &= (\lambda + \eta) \approx \lambda. \end{aligned}$$

Now, taking standard parts, we have that $(Tx,x) \ge \lambda > b$. Since $b = \sup_{\|x\|=1} (Tx,x)$, we have a contradiction. ■

LEMMA 23. $E(b) = I$.

PROOF. Let $\kappa_n > b$, $\kappa_n \to b$ (e.g., $\kappa_n = b(1 + 1/n)$.) By Lemma 22, $E(\kappa_n)x = x$ for all n and all $x \in H$. Hence by Lemma 21 (or even Lemma 19), $E(b)x = x$ for all x. ■

LEMMA 24. If $\lambda < a$, then $E(I) = 0$.

PROOF. Suppose that $\lambda < a$ but $E(\lambda) \neq 0$. Then for some closed subspace $F \neq \{0\}$, $E(\lambda) = P_F$. Let $x \in F$ with $\|x\| = 1$. Choose an appropriate positive $\eta \approx 0$ such that $E(\lambda) = {}^\circ S(\lambda + \eta)$. Set $W = \{i | \lambda_i < \lambda + \eta\}$. Since $x \in F$,

$$\begin{aligned} x &= E(\lambda)x \\ &\approx S(\lambda + \eta)x \\ &= \sum_{\lambda_i < \lambda + \eta} P_i x \\ &= \sum_{i \in W} P_i x \end{aligned}$$

Thus $x \approx \sum_{i \in W} P_i x$.

Since x is standard,

$$\begin{aligned} Tx &= T'x \\ &\approx T'\left(\sum_{i \in W} P_i x\right) \\ &= \sum_{j=1}^{\nu} \lambda_j P_j \left(\sum_{i \in W} P_i x\right) \\ &= \sum_{j \in W} \lambda_j P_j x. \end{aligned}$$

Thus

$$\begin{aligned} (Tx,x) &\approx \left(\sum_{i \in W} \lambda_i P_i x, \sum_{i \in W} P_i x\right) \\ &= \sum_{i \in W} \lambda_i \|P_i x\|^2 \\ &< (\lambda + \eta) \sum_{i \in W} \|P_i x\|^2 \\ &= (\lambda + \eta) \left\|\sum_{i \in W} P_i x\right\|^2 \\ &\approx (\lambda + \eta)\|x\|^2 = (\lambda + \eta) \approx \lambda. \end{aligned}$$

Thus taking standard parts, we obtain

$$(Tx,x) \leq \lambda < a.$$

Since $\|x\| = 1$, this contradicts the fact that $a = \inf_{\|x\|=1} (Tx,x)$. ■

The last three lemmas give (4) of the Spectral Theorem. We conclude with the

PROOF OF (5). Let $\varepsilon > 0$ be real. Partition the interval $[a - h,b]$ as follows:

$$a - h = u_0 < u_1 < u_2 < \cdots < u_k = b,$$

where $u_{i+1} - u_i < \varepsilon/2$. Choose interior points s_i, $u_i \leq s_i \leq u_{i+1}$. The corresponding Riemann–Stieltjes sum is

$$S = \sum_{i=0}^{k-1} s_i[E(u_{i+1}) - E(u_i)].$$

To complete the proof it is only necessary to show that $\|S - T\| < \varepsilon$.

Now in calculating $S - T$ we need to make use of the spectral resolution of T' in terms of the eigenvalues $\{\lambda_i | 1 \leq i \leq \nu\}$, and as we have seen, these eigenvalues are all found in the interval $^*J = {}^*[-\|T\|,\|T\|]$. By Lemma 1, the interval $[a - h,b]$ is contained in the interval $[-\|T\| - h,\|T\|]$. Hence we extend the partition $\{u_i | 0 \leq i \leq k\}$ of $[a - h,b]$ to the partition

$$-\|T\| - h = t_0 < t_1 < \cdots < t_n = \|T\|$$

by including additional points as needed to the left of $a - h$ and to the right of b in such a way that $t_{j+1} - t_j < \varepsilon/2$ for $j = 1,2,\ldots,n$. We write $t_j \leq s_j \leq t_{j+1}, j = 0,1,\ldots,n - 1$, where in case $t_j = u_i$ and $t_{j+1} = u_{i+1}$, then $s_j = r_i$. Then

$$\sum_{j=0}^{n-1} s_j[E(t_{j+1}) - E(t_j)] = \sum_{j=0}^{p} s_j[E(t_{j+1}) - E(t_j)] + \sum_{i=0}^{k-1} r_i[E(u_{i+1}) - E(u_i)] + \sum_{j=q}^{n-1} s_j[E(t_{j+1}) - E(t_j)],$$

where $t_{p+1} = a - h$ and $t_q = b$. But by Lemma 24,

$$\sum_{j=0}^{p} s_j(E(t_{j+1}) - E(t_j)) = \sum_{j=0}^{p} s_j(0 - 0) = 0,$$

and by Lemmas 22 and 23,

$$\sum_{j=q+1}^{n-1} s_j(E(t_{j+1}) - E(t_j)) = \sum_{j=q+1}^{n-1} s_j(I - I) = 0.$$

Hence

$$S = \sum_{j=0}^{n-1} s_j\{E(t_{j+1}) - E(t_j)\}.$$

Let $x \in H$ with $\|x\| = 1$. Let $\eta \approx 0$ be positive such that

$$E(t_j)x \approx S(t_j + \eta)x \quad \text{for} \quad j = 0,1, \ldots ,n.$$

Let

$$W_j = \{i | t_j + \eta < \lambda_i \leq t_{j+1} + \eta\},$$

$j = 0,1, \ldots ,n - 1$. Then $\bigcup_{j=0}^{n-1} W_j = \{i \in {}^*N | 1 < i \leq \nu\}$, and

$$[E(t_{j+1}) - E(t_j)]x \approx s(t_{j+1} + \eta)x - S(t_j + \eta)x = \sum_{i \in W_j} P_i x.$$

Also,

$$x = \sum_{i=1}^{\nu} P_i x = \sum_{j=0}^{n-1} \sum_{i \in W_j} P_i x,$$

$$Tx = \sum_{i=1}^{\nu} \lambda_i P_i x = \sum_{j=0}^{n-1} \sum_{i \in W_j} \lambda_i P_i x.$$

Now

$$\begin{aligned} \|Sx - Tx\| &= \left\| \sum_{j=0}^{n-1} s_j [E(t_{j+1}) - E(t_j)]x - Tx \right\| \\ &\approx \left\| \sum_{j=0}^{n-1} s_j \sum_{i \in W_j} P_i x - \sum_{j=0}^{n-1} \sum_{i \in W_j} \lambda_i P_i x \right\| \\ &= \left\| \sum_{j=0}^{n-1} \left[s_j \sum_{i \in W_j} P_i x - \sum_{i \in W_j} \lambda_i P_i x \right] \right\| \\ &\approx \left\| \sum_{j=0}^{n-1} \sum_{i \in W_j} (s_j - \lambda_i) P_i x \right\|. \end{aligned}$$

Let $\mu_i = s_j - \lambda_i$, where the s_j is selected with $i \in W_j$. Since $t_j \leq s_j \leq t_{j+1}$ and $t_j \leq \lambda_i \leq t_{j+1} + \eta$, we have $|\mu_i| < \varepsilon/2 + \eta$. Thus

$$\begin{aligned} \|Sx - Tx\|^2 &= \left\| \sum_{i=1}^{\nu} \mu_i P_i x \right\|^2 \\ &= \sum_{i=1}^{\nu} |\mu_i|^2 \cdot \|P_i x\|^2 \\ &< \left(\frac{\varepsilon}{2} + \eta \right)^2 \sum_{i=1}^{\nu} P_i(x)^2 \\ &= \left(\frac{\varepsilon}{2} + \eta \right)^2 x^2 = \left(\frac{\varepsilon}{2} + \eta \right)^2. \end{aligned}$$

So,

$$\|Sx - Tx\| < \frac{\varepsilon}{2} + \eta < \tfrac{3}{4}\varepsilon.$$

Finally, since x was an arbitrary point of H with $\|x\| = 1$, we have

$$\|Sx - Tx\| = \sup_{\|x\|=1} \|Sx - Tx\| \leq \tfrac{3}{4}\varepsilon < \varepsilon. \quad \blacksquare$$

REFERENCES

The brief list below is naturally nothing like a complete bibliography of the field. Interested readers are referred to the bibliographies from the papers in [7], [9], and [10]. *Added in Proof*: A comprehensive bibliography has been prepared by D. Randolph Johnson and is available under the title *Lecture Notes in Mathematics No.* 3 from the Department of Mathematics and Statistics of the University of Pittsburgh. Readers will also be interested in H. Jerome Keisler's *Elementary Calculus* with its accompanying teacher's guide *Foundations of Infinitesimal Calculus*, both Prindle, Weber & Schmidt (1976), and in K. D. Stroyan and W. A. J. Luxemburg's *Introduction to the Theory of Infinitesimals*, Academic Press (1976).

1. Bell, J. L. and A. B. Slomson, *Models and Ultraproducts*, North-Holland (1969).
2. Bernstein, Allen R. The Spectral Theorem—a Non-Standard Approach, *Z. Math. Logik Grundlagen Math.*, **18** (1972), pp. 419–434.
3. Bernstein, A. R., Non-Standard Analysis, in *Studies in Model Theory*, M. D. Morley, Ed. *Studies in Mathematics*, Vol. 8 (1973), pp. 35–58.
4. Halmos, Paul R., *Measure Theory*, Van Nostrand (1950).
5. Halmos, Paul R., *Finite-Dimensional Vector Spaces*, 2nd edit., Van Nostrand (1958).
6. Hausner, Melvin, On a Non-Standard Construction of Haar Measure, Comm. Pure Appl. Math., **25** (1972), pp. 403–405.
7. Hurd, Albert (Ed.), *Victoria Symposium on Non-Standard Analysis*, *Lecture Notes in Mathematics*, No. 369, Springer-Verlag (1974).
8. Luxemburg, W. A. J., *Non-Standard Analysis*, *Lectures on Robinson's Theory of Infinitesimals and Infinitely Large Numbers*, Pasadena (1962) and revised edition (1964).
9. Luxemburg, W. A. J. (Ed.), *Applications of Model Theory to Algebra, Analaysis and Probability Theory*, Proceedings of an Internal Symposium on Nonstandard analysis, Holt-Rinehart and Winston, (1969).
10. Luxemburg, W. A. J. (Ed.), *Contributions to Non-Standard Analysis*, Symposium at Oberwolfach, 1970, North-Holland (1972).
11. Luxemburg, W. A. J., What is Nonstandard Analysis?, in *Papers in the Foundations of Mathematics*, *Amer. Math. Monthly*, **80** (1973), No. 6, part II, pp. 38–67.
12. Machover, Moshé and Joram Hirschfeld, *Lectures on Non-Standard Analysis*, *Lecture Notes in Mathematics*, No. 94, Springer-Verlag (1969).
13. Parikh, Robert, A Nonstandard Theory of Topological Groups, in Luxemburg [9], pp. 279–284.
14. Robinson, Abraham, *Non-Standard Analysis*, *Studies in Logic and the Foundations of Mathematics*, North-Holland (1966).
15. Robinson, Abraham, Nonstandard Theory of Dedekind Rings, *Indag. Math.* **29** (1967), pp. 444–452.
16. Robinson, Abraham and Elias Zakon, A Set-Theoretical Characterization of Enlargements, in Luxemburg [9], pp. 109–122.

INDEX

Boldface type is used in this index for the names of certain key theorems and for the numbers of the pages where these theorems are first stated.